LECTURES ON LIE GROUPS

LECTURES ON LIE GROUPS

J. FRANK ADAMS

THE UNIVERSITY OF CHICAGO PRESS
CHICAGO AND LONDON

The University of Chicago Press, Chicago 60637
The University of Chicago Press, Ltd., London

Midway reprint 1982
Printed in the United States of America

LCN: 82-51014
ISBN: 0-226-00530-5

CONTENTS

Page

FOREWORD ix

Chapter

1. BASIC DEFINITIONS 1

2. ONE-PARAMETER SUBGROUPS, THE EXPONENTIAL MAP, ETC. 7

3. ELEMENTARY REPRESENTATION THEORY 22

4. MAXIMAL TORI IN LIE GROUPS 79

5. GEOMETRY OF THE STIEFEL DIAGRAM 101

6. REPRESENTATION THEORY 142

7. REPRESENTATIONS OF THE CLASSICAL GROUPS 165

REFERENCES 180

FOREWORD

These notes derive from a course on the representation-theory of compact Lie groups which I gave in the University of Manchester in 1965, and in particular from duplicated notes on that course which were prepared by Dr. Michael Mather.

It may be asked why one who is not an expert on Lie groups should release such a course for publication. The answer lies partly in the very limited and modest aims of the course; and partly, too, in the continued demand for the duplicated notes, which seems to show that a number of readers sympathise with these aims. I feel that the representation-theory of compact Lie groups is a beautiful, satisfying and essentially simple chapter of mathematics, and that there is a basic minimum of it which deserves to be known to mathematicians of many kinds. In my original lectures I addressed myself mainly to algebraic topologists. If an algebraic topologist tries to read, for example, Borel and Hirzebruch's paper "Characteristic Classes and Homogeneous Spaces" [3] he finds that he needs to know the basic facts about maximal tori, weights and roots of Lie groups. If he tries to read, for example, Bott's "Lectures on K(X)" [4] he finds that he needs to know two main theorems on the representation-theory of compact Lie groups [4, p. 50, Theorem 1; p. 51, Theorem 2]. These theorems appear "in modern dress", but they go back to H. Weyl [22]. I have given these examples for illustration, but they are fairly typical; and they help to indicate a basic syllabus on Lie groups which may be useful to students of many different specialities, from functional analysis and differential geometry to algebra. The object of these notes is to cover this basic syllabus, with proofs, in a reasonably concise way. The material on maximal tori, weights and roots appears in Chapters 4 and 5. The two theorems on representation-theory appear in Chapter 6 as Theorems 6.20 and 6.41. The first three chapters allow one to start the proofs more or less from the beginning.

There is little or no claim to originality; I have simply tried to assemble those lines of argument which I found most attractive in the classical sources. There are perhaps a few small exceptions to this.

(i) In Chapter 3, on elementary representation-theory, I have proceeded in an invariant and coordinate-free way even at certain points where it is usual not to do so. Here my starting-point was a suggestion by H. B. Shutrick for proving the orthogonality relations for characters without first proving the orthogonality relations for the components of a matrix representation (see 3.33(ii) and 3.34(i) below).

Unfortunately, the usual proof of the completeness of characters, following Peter and Weyl [15], makes use of the orthogonality relations for the components of a matrix representation. I was therefore forced to rewrite this also in an invariant way (see 3.46 and 3.47 below).

I have not seen these "invariant" proofs in the sources I have consulted, but I would be sorry to think they were not known to the experts.

(ii) In the same chapter, I have laid particular stress on real and symplectic representations, which are important to topologists; and I have preferred those methods which apply simultaneously to the real and to the symplectic case.

(iii) Theorem 5.47 allows one to read off the fundamental group of a compact connected group from its Stiefel diagram; the statement is surely well known to the experts, and is undoubtedly implicit in Stiefel's work, but I do not remember seeing an explicit statement or proof in the sources I have consulted.

(iv) It is usual to give a meaning to the words "highest weight" by ordering the weights lexicographically, in a way which is somewhat arbitrary; I have preferred to use instead a partial ordering which is manifestly invariant, and which seems to me to have some technical advantages (see 6.22 and 6.23 below). I hope this departure from tradition may commend itself to other workers.

I am most grateful to A. Borel, to Harish-Chandra and particularly to H. Samelson for giving me tutorials on Lie groups and representation-theory. I have also profited from

R. G. Swan's "Notes on Maximal Tori, etc.". I am also very grateful to Michael Mather, who prepared the notes on the original course. In particular, the trick in the present proof of 2.19 is due to him; it allowed him to slim the original lectures by removing a good deal of standard material on the relation between a Lie group and its Lie algebra. He also removed a good deal of hard work from the proof of 5.55. Finally, I am grateful to H. B. Shutrick for the suggestion noted above.

Chapter 1

BASIC DEFINITIONS

1.1 DEFINITIONS. Let V, W be finite dimensional vector spaces over the real numbers R. Let U be an open subset of V, f a map from U to W, and x a point of U. Then f is *differentiable at x* if there is a linear map $f'(x) : V \to W$ such that

$$f(x + h) = f(x) + (f'(x))(h) + o|h| .$$

If f is differentiable at each point of U, we say that f is *differentiable* on U. In this case we have a function

$$f' : U \to \mathrm{Hom}(V, W),$$

and we may ask if this is differentiable. We say that f is *smooth* (or of *class* C^∞) on U if each function f, f', f'', ... is differentiable on U. (Of course, the definition of each of these depends on the previous one being defined and differentiable.)

1.2 DEFINITIONS. If X is a topological space and V a finite dimensional vector space, a chart is a homeomorphism $\varphi_\alpha : U_\alpha \to X_\alpha$, where $U_\alpha \subset V$ is open and $X_\alpha \subset X$ is open.

An atlas is a collection of charts $\{\varphi_\alpha\}$ with $\cup X_\alpha = X$. The atlas is smooth if the functions $\varphi_\beta^{-1}\varphi_\alpha$, defined on $\varphi_\alpha^{-1}(X_\alpha \cap X_\beta)$, are smooth.

Let X, Y be topological spaces with smooth atlases $\{\varphi_\alpha\}$ and $\{\psi_\beta\}$. Then a map $f : X \to Y$ is smooth if the maps $\psi_\beta^{-1} f \varphi_\alpha$, defined on $\varphi_\alpha^{-1}(X_\alpha \cap f^{-1} Y_\beta)$, are smooth. Notice that the composition of two smooth maps is smooth, and the identity map of a space with atlas is smooth.

Two atlases $\{\varphi_\alpha\}$, $\{\psi_\beta\}$ on X are equivalent if the maps

$$1 : X, \{\varphi_\alpha\} \to X, \{\psi_\beta\}$$

$$1 : X, \{\psi_\beta\} \to X, \{\varphi_\alpha\}$$

are smooth.

A differential or smooth manifold is a Hausdorff space with an equivalence class of smooth atlases. This equivalence class is called its differential structure.

1.3 PROPOSITION. If X, Y are smooth manifolds, then $X \times Y$ can be given the structure of a smooth manifold in a unique way to satisfy:

(i) $\pi_1 : X \times Y \to X$ and $\pi_2 : X \times Y \to Y$ are smooth maps.

(ii) $f : Z \to X \times Y$ is smooth if and only if $\pi_1 f$ and $\pi_2 f$ are smooth.

Proof. Given charts φ_α in X, ψ_β in Y, form the chart $\varphi_\alpha \times \psi_\beta$ in $X \times Y$. Do this for each pair, α, β. The rest of the proof consists of checking the necessary properties, and will be left to the reader.

1.4 DEFINITIONS. A Lie group G is

(i) a smooth manifold, and

(ii) a group, with product $\mu : G \times G \to G$ and inverse $i : G \to G$, such that

(iii) μ and i are smooth.

A homomorphism of Lie groups $\theta : G \to H$ is

(i) a homomorphism of groups, and

(ii) a smooth map.

1.5 EXAMPLES

1. R^n considered as a group under addition, with an atlas of just one chart given by the identity map.

2. $T^n = R^n/Z^n$ (where Z^n is the set of points in R^n all of whose co-ordinates are integers) considered as a quotient

group of R^n, with charts given by the restriction of the projection $R^n \to T^n$ to small open sets.

3. Let V be a finite dimensional vector space over R. Then Aut V, the set of automorphisms of V, is an open subset of Hom(V,V) given by det $\neq 0$, so Aut V is a smooth manifold, and is a group under composition. The product map is smooth, since it is given by polynomials $(\Sigma a_{ij} b_{jk})$ and the inverse map is smooth since it is given by polynomials divided by the determinant. Thus Aut V is a smooth manifold. Aut R^n is called GL(n, R).

This also works over the complex numbers or the quaternions. For instance, $\mathrm{Hom}_C(V, V)$ is a linear subspace of $\mathrm{Hom}_R(V, V)$, and $\mathrm{Aut}_C V = \mathrm{Aut}_R V \cap \mathrm{Hom}_C(V, V)$. So $\mathrm{Aut}_C V$ is an open subset of $\mathrm{Hom}_C(V, V)$.

1.6 DEFINITION of the <u>tangent bundle</u> of a smooth manifold X. Let $\{\varphi_\alpha : U_\alpha \to X_\alpha\}$ be an atlas based on the vector space V. Take the disjoint union of the spaces $X_\alpha \times V$ over all α and, whenever $x \in X_\alpha \cap X_\beta$, identify $(x,v) \in X_\alpha \times V$ with $(x, (\varphi_\beta^{-1}\varphi_\alpha)' v) \in X_\beta \times V$. Call the identification space T(X), and define $p : T(X) \to X$ by projection on the first factor of each product. This is the <u>tangent bundle</u>. It is an invariant of X.

We call $p^{-1}x$ the tangent space at the point $x \in X$, written X_x, and a point of X_x is a tangent vector at x.

Note that $T(X)$ can be made into a smooth manifold in an obvious way, and p is a smooth map.

Given a smooth map $f : X \to Y$ construct a smooth natural bundle map $f_* : T(X) \to T(Y)$ as follows. For $x \in X_\alpha$ and $fx \in Y_\beta$ set $f_*(x, v) = \left(fx, \left(\psi_\beta^{-1} f \varphi_\alpha\right)' v\right)$.

1.7 NOTATION. Let G be a Lie group, with unit e. Then we write $L(G)$ for G_e and $L(f)$ for $f_* | G_e$. Then L is a functor. We also write f' for f_*, in the light of the following example:

1.8 EXAMPLE. Consider example 1.5.1. Then the tangent space at the origin of R^n may be identified with R^n under the chart. If $f : R^m \to R^n$ is a smooth map, then $f_* | R^n_o = f'$ under this identification.

1.9 DEFINITIONS. A smooth vector field on a manifold X is a smooth cross-section of the tangent bundle. That is, it is a smooth map $\lambda : X \to T(X)$ such that $p\lambda = 1$.

Let G be a Lie group. For $x \in G$ define $L_x : G \to G$ by $L_x(g) = xg$. This is smooth. Then a smooth vector field λ on G is left invariant if the following diagram is commutative for each $x \in G$:

$$\begin{array}{ccc} T(G) & \xrightarrow{L_{x*}} & T(G) \\ \uparrow \lambda & & \uparrow \lambda \\ G & \xrightarrow{L_x} & G \end{array}$$

1.10 DEFINITION. Let G be a Lie group. For each $x \in G$ define $A_x : G \to G$ by $A_x(g) = xgx^{-1}$. This is a smooth automorphism, and hence defines a linear map $A'_x : G_e \to G_e$; that is, $A'_x \in \operatorname{Aut} G_e$. Hence $x \to A'_x$ defines a map $\operatorname{Ad} : G \to \operatorname{Aut} G_e$. This is a smooth homomorphism.

Chapter 2

ONE-PARAMETER SUBGROUPS, THE EXPONENTIAL MAP, ETC.

2.1 LEMMA. Suppose given a smooth vector field $v(x)$ defined in a neighbourhood U of 0 in R^n. Consider the following equations for a function $f : R^1 \to R^n$, namely, $f'(t, 1) = v(f(t))$, $f(0) = 0$. Then there is $\epsilon > 0$ for which the equations have a solution in $(-\epsilon, \epsilon)$, and this solution is both unique and smooth.

This is a particular case of the more general:

2.2 LEMMA. Let $U \subset R^n$ and $V \subset R^m$ be neighbourhoods of 0, y_o respectively. Let $v(x, y)$ be a vector field in R^n depending smoothly on $x \in U$ and $y \in V$. Consider the equations $f(0) = 0$, $f'(t, 1) = v(f(t), y)$, for each fixed $y \in V$, as equations for a function $f : R^1 \to R^n$. Then there is $\epsilon > 0$ and a neighbourhood V' of y_o in R^m such that a solution exists in $(-\epsilon, \epsilon)$ for each $y \in V'$, this solution is unique, and depends smoothly on $t \in (-\epsilon, \epsilon)$ and $y \in V'$.

Proof. We refer the reader to [12, p. 94, Proposition 1], [5, Chapter 2, Theorem 4.1], [2, Appendix, Section II], or [8, Chapter 9, Theorem 1].

2.3 DEFINITION. A 1-parameter subgroup of G is a homomorphism of Lie groups $\theta : R^1 \to G$, where R^1 is a Lie group under addition with an atlas with one chart given by the identity map.

2.4 EXAMPLE. In $T^2 = R^2/Z^2$ set $\theta(t) = (t, ct)$ for c any constant.

2.5 Let θ be a 1-parameter subgroup of G. Let (0, 1) be the unit tangent vector at the origin in R^1. Associate with θ the vector $\theta'(0, 1) \in G_e$. Then:

2.6 THEOREM. This sets up a 1-1 correspondence between 1-parameter subgroups of G and vectors in G_e.

Proof. We need:

2.7 LEMMA. Let X be a smooth manifold, v(x) a smooth

vector field on X, and θ, $\varphi : [a, b] \to X$ two smooth functions satisfying $\theta'(t, 1) = v(\theta(t))$

$$\varphi'(t, 1) = v(\varphi(t))$$

$$\theta(a) = \varphi(a).$$

Then $\theta(t) = \varphi(t)$ for all $t \in [a, b]$.

Proof. Let c be the least upper bound of the set of d for which $\theta(t) = \varphi(t)$ on $[a, d]$. Then $\theta(c) = \varphi(c)$ by continuity. If $c < b$ we may take local coordinates at $\theta(c) = \varphi(c)$ and apply 2.1, showing that $\theta(t) = \varphi(t)$ in some $(c - \epsilon, c + \epsilon)$, which contradicts the definition of c. Thus $c = b$.

Proof of 2.6.

(i) Uniqueness. Suppose that θ corresponds to $v \in G_e$. The vector (0, 1) can be extended to a left invariant vector field (t, 1) on R^1, and v can be extended to a left invariant vector field v(x) on G. Taking the diagram of tangent spaces corresponding to

$$\begin{array}{ccc} R & \xrightarrow{\theta} & G \\ {\scriptstyle L_t}\downarrow & & \downarrow{\scriptstyle L_{\theta(t)}} \\ R & \xrightarrow{\theta} & G \end{array}$$

we see that $\theta'(t, 1) = L'_{\theta(t)} v = v(\theta(t))$. Thus, by 2.7, θ is unique.

(ii) Existence. Given $v \in G_e$, extend v to a left invariant vector field $v(x)$ on G. Then the equations $\theta'(t, 1) = v(\theta(t))$, $\theta(0) = 0$ have a solution for $t \in (-\epsilon, \epsilon)$, by 2.1.

We will show, firstly, that $\theta(s)\theta(t) = \theta(s + t)$ for $|s| < \frac{1}{2}\epsilon$, $|t| < \frac{1}{2}\epsilon$. Well, for s fixed, $\theta(s)\theta(t)$ and $\theta(s + t)$ are both solutions of $\varphi'(t, 1) = v(\varphi(t))$, $\varphi(0) = \theta(s)$. Thus, by 2.7, $\theta(s)\theta(t) = \theta(s + t)$.

Now define $\psi : R^1 \to G$ as follows. For $t \in R^1$ choose a positive integer N such that $|\frac{t}{N}| < \frac{\epsilon}{2}$, and set $\psi(t) = \left(\theta\left(\frac{t}{N}\right)\right)^N$. Then ψ is well-defined since, if M is another such integer, $\left(\theta\left(\frac{t}{MN}\right)\right)^N = \theta\left(\frac{t}{M}\right)$, by the previous paragraph, so $\left(\theta\left(\frac{t}{N}\right)\right)^N = \left(\theta\left(\frac{t}{MN}\right)\right)^{MN} = \left(\theta\left(\frac{t}{M}\right)\right)^M$. Further, ψ is a group homomorphism, for, if $|\frac{s}{N}| < \frac{\epsilon}{2}$ and $|\frac{t}{N}| < \frac{\epsilon}{2}$ we have

$$\psi(s + t) = \left(\theta\left(\frac{s + t}{N}\right)\right)^N = \left(\theta\left(\frac{s}{N}\right)\right)^N \left(\theta\left(\frac{t}{N}\right)\right)^N = \psi(s)\psi(t).$$

Now ψ is also smooth, and extends θ. So ψ is a 1-parameter subgroup and $\psi'(0, 1) = v$.

2.8 DEFINITION of the <u>exponential map</u>. Define $\exp : G_e \to G$ as follows. Let $v \in G_e$ and let θ_v be the corresponding 1-parameter subgroup of G. Then $\exp(tv) = \theta_v(t)$.

We need to show that $\theta_v(t)$ depends only on tv. Well, for fixed s, $\theta_v(st)$ is clearly the 1-parameter subgroup corresponding to sv. Thus $\theta_v(st) = \theta_{sv}(t)$, so $\theta_v(s) = \theta_{sv}(1)$.

2.9 THEOREM. exp is smooth.

Proof. Let $v_o \in G_e$. We show that exp is a smooth function in a neighbourhood of v_o.

Well, $\theta_v(t)$ is the solution of the differential equation

$$\theta_v'(t, 1) = v(\theta_v(t))$$
$$= L'_{\theta_v(t)} v.$$

Now $L'_x v$ is a smooth function of $x \in G$ and $v \in G_e$. So (2.2) the solution is a smooth function of t and v for $|t| < \epsilon$ and v in a neighbourhood of v_o.

Take a positive integer N with $\frac{1}{N} < \epsilon$. Then $\exp v = \theta_v(1) = \left(\theta_v\left(\frac{1}{N}\right)\right)^N$, which is a smooth function of v in a neighbourhood of v_o.

2.10 REMARK. $\exp : G_e \to G$ induces $\exp' = 1 : G_e \to G_e$ and

2.11 PROPOSITION. exp is natural. That is, given a homomorphism of Lie groups $\varphi : G \to H$ inducing $\varphi' : G_e \to H_e$, the following diagram is commutative:

$$\begin{array}{ccc} G_e & \xrightarrow{\varphi'} & H_e \\ {\scriptstyle \exp}\downarrow & & \downarrow{\scriptstyle \exp} \\ G & \xrightarrow{\varphi} & H \end{array}$$

Proof. Let $v \in G_e$, and let $\theta : R^1 \to G$ be the corresponding 1-parameter subgroup of G. Then $\varphi\theta : R^1 \to H$ is the 1-parameter subgroup of H corresponding to $\varphi' v$, since the derivative is natural. Thus $\exp \varphi' v = \varphi\theta(1) = \varphi \exp v$.

2.12 EXAMPLE. Let V be a finite dimensional real vector space, and take $G = \mathrm{Aut}\, V$, which is an open subspace of $\mathrm{Hom}(V, V)$. We can identify G_e with $\mathrm{Hom}(V, V)$. Let $A \in \mathrm{Hom}(V, V)$. Then we assert

$$\exp A = 1 + A + \frac{A^2}{2} + \dots + \frac{A^n}{n!} + \dots .$$

Proof. Consider $1 + At + \frac{A^2t^2}{2} + \dots + \frac{A^nt^n}{n!} + \dots .$ This is easily seen to be a smooth homomorphism from R^1 to Aut V, and is the 1-parameter subgroup corresponding to A. Thus

$$\exp A = 1 + A + \frac{A^2}{2} + \dots + \frac{A^n}{A!} + \dots .$$

2.13 EXAMPLE. Consider $G = T^n = R^n/Z^n$. Then $G_e = R^n$, and exp can be identified with the covering map $R^n \to T^n$.

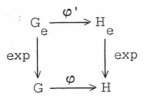

2.14 THEOREM. exp is a diffeomorphism of a neighbourhood of $0 \in G_e$ with a neighbourhood of e in G.

Proof. This is immediate from 2.10 and the Jacobian theorem. (See [12, p. 12, Theorem 1].)

2.15 THEOREM. Let $G_e = V_1 \oplus V_2$, and define $\varphi : G_e \to G$ by $\varphi(v_1, v_2) = \exp v_1 \exp v_2$. Then φ is a diffeomorphism of a neighbourhood of $0 \in G_e$ with a neighbourhood of e in G.

Proof. φ is the composition $V_1 \oplus V_2 \xrightarrow{\exp \times \exp} G \times G \xrightarrow{\mu} G$, and so is differentiable. Further, φ' is the identity on both V_1 and V_2, and so is the identity on G_e. We may proceed as in 2.14.

2.16 PROPOSITION. Let G_1 denote the identity component of G, and let $S \subset G_1$ be a neighbourhood of e. Then the subgroup generated by S is G_1.

Proof. Clearly $gp\{S\} \subset G_1$. Now $gp\{S\}$ is an open subgroup of G_1, so all its cosets are open. Thus $gp\{S\}$ is also closed, so $gp\{S\} = G_1$.

2.17 THEOREM. If G is connected, a homomorphism of Lie

groups $\theta : G \to H$ is determined by the induced homomorphism $\theta' : G_e \to H_e$.

Proof. By 2.11 we have the commutative diagram:

$$\begin{array}{ccc} G_e & \xrightarrow{\theta'} & H_e \\ {\scriptstyle \exp}\downarrow & & \downarrow{\scriptstyle \exp} \\ G & \xrightarrow{\theta} & H \end{array}$$

Thus θ is determined by θ' at least on the subgroup of G generated by the image of exp. But this is a neighbourhood of e in G, so θ is determined on G.

2.18 LEMMA. Let $\varphi_\alpha : U_\alpha \to G_\alpha$ be a chart on G which sends $0 \in V$ to $e \in G$. Then, omitting φ_α, we can write $xy = x + y + o(r)$ in a neighbourhood of e in G, where $r = r(x, y)$ denotes the distance of (x, y) from (e, e) in $G \times G$ under a metric.

Proof. Since the product in G is differentiable, there is a constant vector a and constant linear functions b, c such that $xy = a + bx + cy + o(r)$. Set $x = e$ and we find that $y = a + cy + o(r)$ so $a = 0$, $c = 1$. Similarly $b = 1$, so $xy = x + y + o(r)$.

2.19 THEOREM. A connected Abelian Lie group G has the form $T^a \times R^b$.

Proof. We show first that $\exp : G_e \to G$ is a homomorphism. Well,

$$\exp s \exp t = \left(\exp \frac{s}{N}\right)^N \left(\exp \frac{t}{N}\right)^N$$

$$= \left(\exp \frac{s}{N} \exp \frac{t}{N}\right)^N$$

since G is Abelian

$$= \left[\exp\left(\frac{s}{N} + \frac{t}{N} + o\left(\frac{1}{N}\right)\right)\right]^N$$

by 2.18 and 2.14, where we consider s, t fixed and N varying

$$= \exp(s + t + o(1))$$

$$= \exp(s + t).$$

Thus exp is a homomorphism, and, by 2.16, exp is onto.

Consider K = Ker exp. By 2.14, since exp is a homomorphism, K is discrete. Now a discrete subgroup of a real vector space is a free Abelian group, with generators $g_1, \ldots, g_r$ which are linearly independent over R. (This is proved by induction over the dimension of the vector space.) Extend this to a basis of G_e. Then K is expressed as the set of points with coordinates $(n_1, \ldots, n_r, 0, \ldots, 0)$, each $n_i \in Z$. Thus $G = G_e/K = T^r \times R^{n-r}$.

2.20 COROLLARY. A Lie group which is compact, connected and Abelian is a torus.

2.21 EXERCISE. Classify the compact Abelian Lie groups.

2.22 DEFINITION of submanifold.

Let $W \subset V$ be a real finite dimensional vector space and subspace. Let M be a smooth manifold, and N a subset of M. Then a chart $\varphi_a : V_a \to M_a$ is good if

(i) $M_a \cap N = \emptyset$ (the empty set), or

(ii) φ_a sends $V_a \cap W$ onto $M_a \cap N$.

An atlas is good if all its charts are good.

N is a submanifold if there is a good atlas in the differential structure of M.

Equivalence of good atlases is defined by the identity map being smooth, as before.

2.23 PROPOSITION. If N is a submanifold of M, then N can be given a differential structure as a manifold so that

(i) the inclusion of N in M is smooth, and

(ii) $P \to N$ is smooth if and only if the composition $P \to N \to M$ is smooth.

Proof is clear, and left to the reader.

2.24 REMARK. It follows that $T(N)$ is embedded in $T(M)$.

2.25 EXERCISE. If N_1, N_2 are submanifolds of M_1, M_2 then $N_1 \times N_2$ is a submanifold of $M_1 \times M_2$, and its differential structure as a submanifold is the same as its differential structure as a product.

2.26 PROPOSITION. If G is a Lie group, and H is both a submanifold and a subgroup, then H is a Lie group.

Proof. Apply 2.23 and 2.25 to the maps of pairs $\mu : G \times G, H \times H \to G, H$ and $i : G, H \to G, H$.

2.27 THEOREM. A closed subgroup H of a Lie group G is a submanifold.

Proof. The next three lemmas constitute a proof.

2.28 LEMMA. In 2.27, suppose G_e has a norm. Suppose $0 \neq h_n \in G_e$ is a sequence of points such that $\exp h_n \in H$, $h_n \to 0$, and $\frac{1}{|h_n|} h_n \to v \in G_e$. Then $\exp(tv) \in H$ for all $t \in R$.

Proof. $\frac{t}{|h_n|} h_n \to tv$, and $|h_n| \to 0$, so we may choose integers m_n such that $m_n |h_n| \to t$. Then $\exp m_n h_n \to \exp tv$. But $\exp m_n h_n = (\exp h_n)^{m_n} \in H$, and H is closed. So $\exp tv \in H$, as required.

Let W be the set of such tw in G_e. Then $\exp W \subset H$.

2.29 LEMMA. W is a vector subspace of G_e.

Proof. Clearly $w \in W$ implies $tw \in W$ all $t \in R$.

So, suppose $w_1, w_2 \in W$, and suppose $w_1 + w_2 \neq 0$. We will show that $w_1 + w_2 \in W$.

Consider $\exp(tw_1)\exp(tw_2)$. This is in H. For t sufficiently small we can write $\exp(tw_1)\exp(tw_2) = \exp(f(t))$, where $f(t)$ is a smooth curve in G_e and $f(0) = 0$.

Now $\exp(tw_1)\exp(tw_2) - \exp t(w_1 + w_2) = o(t)$, by 2.18, so $\frac{1}{t}f(t) \to w_1 + w_2$ as $t \to 0$. Thus we may apply 2.28 with $h_n = f\left(\frac{1}{n}\right)$, for n sufficiently large, and $v = \frac{1}{|w_1 + w_2|}(w_1 + w_2)$, and deduce that $w_1 + w_2 \in W$.

2.30 LEMMA. exp W is a neighbourhood of e in H.

Proof. Split G_e as $W' \oplus W$ and consider the diffeomorphism $\varphi(w', w) = \exp(w')\exp(w)$ between a neighbourhood of 0 in G_e

and a neighbourhood of e in G (2.15). Suppose the lemma does not hold. Then there is a sequence of pairs (w'_n, w_n) such that $\exp(w'_n)\exp(w_n) \in H$, $\exp(w'_n)\exp(w_n) \to e$, and $w'_n \neq 0$. Since $\exp(w_n) \in H, \exp(w'_n) \in H$. Then we can find a subsequence of w'_n such that $\frac{1}{|w'_n|} w'_n \to w' \in W'$, for some such w', and $|w'| = 1$. It follows from 2.28 that $w' \in W$, which is a contradiction.

Thus exp W is a neighbourhood of e in H.

It is now clear that exp provides a good chart for a neighbourhood of e. Left translation gives a good chart round any other point of H. This completes the proof of 2.27.

2.31 EXAMPLES.

$O(n) \subset GL(n, R)$

$U(n) \subset GL(n, C)$

$SP(n) \subset GL(n, Q)$

are closed subgroups, and so submanifolds, of Lie groups. Thus they are Lie groups.

In each case, the tangent space at e of the subgroup consists of the matrices X such that $\bar{X}^T = -X$.

Proof.

(i) Suppose X is in the tangent space at e of the subgroup. Take a smooth curve of the form $f(t) = 1 + tX + o(t)$ in the subgroup. Then $\overline{f(t)}^T f(t) = 1$, by definition of the subgroup. That is,

$$(1 + t\bar{X}^T + o(t))(1 + tX + o(t)) = 1.$$

Thus

$$\bar{X}^T + X = 0.$$

(ii) Suppose $\bar{X}^T = -X$.

Then

$$(\overline{\exp tX})^T = \left(\overline{\Sigma_0^\infty t^n X^n/n!}\right)^T \qquad \text{by 2.12}$$

$$= \Sigma t^n (\bar{X}^T)^n/n!$$

$$= \Sigma t^n (-X)^n/n!$$

$$= (\exp tX)^{-1}.$$

Therefore $\exp(tX)$ lies in the subgroup, and X in the tangent space.

2.32 EXAMPLE. If G is a compact Lie group, and H is a closed connected Abelian subgroup, then H is a torus.

2.33 PROPOSITION. A closed connected subgroup H of a Lie group G is determined by its tangent space at e.

Proof. See 2.17.

2.34 DEFINITION. Suppose G is a Lie group and H a closed subgroup. Then the <u>quotient space</u> G/H is the set of cosets gH. We have the projection $p : G \to G/H$, and give G/H the quotient topology.

2.35 EXERCISE. G/H is Hausdorff.

2.36 PROPOSITION. If H is a closed subgroup of a Lie group G, we can give G/H a differential structure as a manifold so that

(i) p is smooth,

(ii) $f : G/H \to M$ is smooth if and only if $fp : G \to M$ is smooth.

<u>Proof</u>. Split G_e as $W' \oplus W$ where $W = H_e$ as before. Let U be a small neighbourhood of 0 in W', and define $\psi : U \to G/H$ by $U \to W' \to G_e \xrightarrow{\exp} G \xrightarrow{p} G/H$. Then ψ is a homeomorphism with a neighbourhood of eH. The rest of the proof is left as an exercise for the reader.

2.37 PROPOSITION. $H \to G \to G/H$ is a fibration.

<u>Proof</u>. See [17, I.7.5].

Chapter 3

ELEMENTARY REPRESENTATION THEORY

In this chapter we set up elementary representation-theory. The basic definitions and constructions occupy 3.1 to 3.13. Then we introduce integration. From this we draw the usual consequences, including complete reducibility (3.15 to 3.21). Then comes Schur's Lemma, some of its consequences, and the definition of the representation ring (3.22 to 3.28). Then traces, characters and the orthogonality relations (3.29 to 3.37). Then the Peter-Weyl theorem and the completeness of characters (3.38 to 3.49). Then the usual material on real and symplectic representations (3.50 to 3.64). Next comes the behaviour of the representation ring for products (3.65 to 3.67) and coverings (3.68 to 3.70). Finally we have the representation-theory of the torus (3.71 to 3.78).

3.1 DEFINITIONS. Let Λ be one of the classical fields R (the real numbers), C (the complex numbers) or Q (the quaternions). Let G be a topological group. Then a _ΛG-space_ is a finite-dimensional vector space V over Λ provided with a continuous homomorphism

$$\theta : G \to \mathrm{Aut}\ V.$$

(Such a V is also called a _representation_ of G over Λ or a _G-space_ over Λ.)

Alternatively, for each $g \in G$ and $v \in V$ we are given $gv \in V$, and the following conditions are satisfied.

(i) $ev = v$ and $g(g'v) = (gg')v$.

(ii) gv is a Λ-linear function of v.

(iii) gv is a continuous function of g and v.

By choosing a base in V we can regard θ as taking values in $GL(n, \Lambda)$. We then speak of a _matrix representation_. In the case $\Lambda = Q$, if we wish to write our matrices on the left, it will be prudent to arrange that V is a right module over Q. Fortunately we can make any left module over Q into a right module over Q, and vice versa, by the formula

$$qv = v\bar{q} \quad (q \in Q,\ v \in V).$$

Here the conjugate of a quaternion is defined as usual: if

$q = a + bi + cj + dk$, then $\bar{q} = a - bi - ij - dk$.

Let V and W be ΛG-spaces. A G-map is a function $f : V \to W$ which commutes with the action of G, that is,

$$f(gv) = g(fv).$$

A ΛG-map is a G-map which is Λ-linear; mostly we deal with such. The set of such ΛG-maps is written $\mathrm{Hom}_{\Lambda G}(V, W)$, or sometimes simply $\mathrm{Hom}_G(V, W)$ if Λ is understood. It is a vector space over R if $\Lambda = R$ or Q, over C if $\Lambda = C$.

A ΛG-isomorphism is a ΛG-map which has an inverse. As usual, we say that two ΛG-spaces are equivalent if they are isomorphic.

3.2 DEFINITION. Let V be a G-space over C. A structure map on V is a G-map $j : V \to V$ such that

(i) j is conjugate-linear, that is,

$$j(zv) = \bar{z}(jv) \quad (z \in C), \quad \text{and}$$

(ii) $j^2 = \pm 1$.

3.3 EXPLANATION. If V is a G-space over Q, we may regard it as a G-space over C with a structure map such that $j^2 = -1$. Actually we may do so in two ways. On the one hand we can take the C-module structure given by i acting on the left and the structure map given by j acting on the left. On the other hand we can take the C-module structure given by i acting on the right (-i acting on the left) and the structure map given by j acting on the right (-j acting on the left). It makes no difference which we take, because we can define an automorphism $\alpha : V \to V$ taking one structure into the other, for example $\alpha(v) = kv$.

Conversely, given a G-space over C with a structure map such that $j^2 = -1$, we can clearly reconstruct a G-space over Q.

Similarly, it is often convenient to regard a G-space V over R as being equivalent to a G-space V' over C provided with a structure map such that $j^2 = +1$. To pass from V to V' we take $V' = C \otimes_R V$ provided with the obvious operations and structure maps:

$$z(z' \otimes v) = zz' \otimes v \quad (z, z' \in C)$$

$$g(z \otimes v) = z \otimes gv$$

$$j(z \otimes v) = \bar{z} \otimes v.$$

To pass from V' to V we split V' into the +1 and -1 eigenspaces of j; these are G-spaces over R which are isomorphic under i. These operations are clearly inverse to one another, up to isomorphism.

3.4 DEFINITION. Given ΛG-spaces V and W, we can form the direct sum of the two vector spaces, $V \oplus W$, and make G act on it by

$$g(v,w) = (gv, gw).$$

Equivalently, we may take two G-spaces V and W over C with structure maps j_V, j_W such that $j_V^2 = j_W^2$, and put on $V \oplus W$ the structure map $j_V \oplus j_W$.

The next five operations start from a G-space V over Λ and construct a G-space over some Λ'. The possibilities are displayed in the following diagram, which is not commutative.

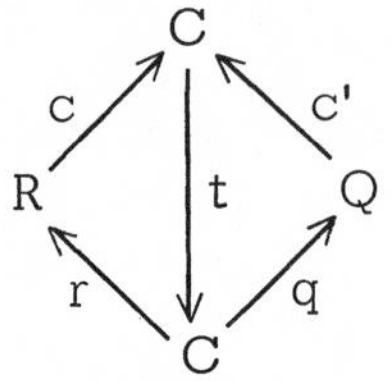

3.5 DEFINITIONS.

(i) If V is a G-space over R, define $cV = C \otimes_R V$, regarded as a G-space over C as in 3.3.

(ii) Similarly, if V is a G-space over C, define $qV = Q \otimes_C V$, and regard it in the obvious way as a G-space and a left module over Q.

(iii) If V is a G-space over Q, let c'V have the same underlying set as V and the same operations from G, but regard it as a vector space over C.

(iv) Similarly, if V is a G-space over C, let rV have the same underlying set as V and the same operations from G, but regard it as a vector space over R.

(v) Let V be a G-space over C. We define tV to have the same underlying set as V and the same operations from G, but we make C act in a new way: z acts on tV as $\bar{z}$ used to act on V.

Let us adopt the viewpoint of 3.3; then both c and c' act on a G-space over C provided with a structure map j, and they act by forgetting the structure map.

All these constructions are natural; given a ΛG-map $f : V \to W$, we can construct maps cf, qf, c'f, rf and tf.

All these constructions commute with direct sums $\oplus$.

3.6 PROPOSITION.

rc = 2

cr = 1 + t

qc' = 2

$c'q = 1 + t$

$tc \;\; = c$

$rt \;\; = r$

$tc' \quad c'$

$qt \;\; = q$

$t^2 \;\; = 1.$

These equations are to be interpreted as saying that

$rcV \cong V \oplus V$ for each V over R

$crV \cong V \oplus tV$ for each V over C,

etc.

Proof. Most of this can safely be left to the reader; we show that $cr = 1 + t$.

Let V be a G-space over C. We want to study $C \otimes_R V$ with C acting on the first factor and G on the second. Let C act on the first factor of $C \otimes_R C$, and let $C \otimes_R C$ act on $C \otimes_R V$ in the obvious way. Then $C \otimes_R V$ is a G-space over the C-algebra $C \otimes_R C$.

We will now split the unit $1 \otimes 1$ of $C \otimes_R C$ into orthogonal idempotents, and so obtain a splitting of $C \otimes_R V$. In detail, let

$$e_1 = \frac{1}{2}(1 \otimes 1 + i \otimes i)$$

$$e_2 = \frac{1}{2}(1 \otimes 1 - i \otimes i).$$

Then $e_1^2 = e_1$, $e_2^2 = e_2$, $e_1 e_2 = 0$ and $e_1 + e_2 = 1$, as required. So

$$C \otimes_R V \cong e_1(C \otimes_R V) \oplus e_2(C \otimes_R V),$$

where the isomorphism is a G-isomorphism over C. Further,

$V \cong e_2(C \otimes_R V)$ by, for instance, $v \to e_2(1 \otimes v)$ and

$tV \cong e_1(C \otimes_R V)$ by, for instance, $v \to e_1(1 \otimes v)$.

Thus $crV \cong V \oplus tV$.

3.7 DEFINITION. Given G-spaces V and W over C, we can form the tensor product of the two vector spaces, $V \otimes_C W$, and make G act on it by

$$g(v \otimes w) = gv \otimes gw.$$

Suppose now that V and W admit structure maps j_V, j_W such that $j_V^2 = \epsilon_V$, $j_W^2 = \epsilon_W$. Then $V \otimes_C W$ admits a structure map $j = j_V \otimes j_W$ such that $j^2 = \epsilon_V \epsilon_W$. We can separate three cases.

(i) $\epsilon_V = \epsilon_W = +1$. The tensor product of two real representations is real. The construction amounts to taking two G-spaces V, W over R and forming $V \otimes_R W$.

(ii) $\epsilon_V = +1$, $\epsilon_W = -1$. The tensor product of one real representation V and one quaternionic representation W is quaternionic. The construction amounts to taking $V \otimes_R W$ and making Q act on it by

$$q(v \otimes w) = v \otimes qw.$$

The case $\epsilon_V = -1$, $\epsilon_W = +1$ is similar.

(iii) $\epsilon_V = \epsilon_W = -1$. The tensor product of two quaternionic representations V and W is real. It is natural to interpret the construction in terms of $V \otimes_Q W$. For this to make sense we must consider V as a right module over Q and W as a left module over Q. If we use the resulting structure maps, then the -1 eigenspace of the structure map $j_V \otimes j_W$ on $V \otimes_C W$ coincides with $V \otimes_Q W$.

3.8 PROPOSITION.

(i) All our tensor products are compatible with the maps c, c'.

(ii) The tensor product $\otimes$ is bilinear over the direct sum $\oplus$.

3.9 DEFINITION. Given G-spaces V and W over the same field Λ, we can form $\mathrm{Hom}_\Lambda(V,W)$, the set of Λ-linear maps from V to W. It is a vector space over R if $\Lambda = R$ or Q, over C if $\Lambda = C$. We can make G act on it by

$$(gh)v = g(h(g^{-1}v)) \quad (h \in \mathrm{Hom}_\Lambda(V,W))$$

or equivalently

$$gh = (\theta_W g)h(\theta_V g^{-1}).$$

(Note that $\mathrm{Hom}_\Lambda(V,W)$ is covariant in W and contravariant in V.) The subspace of elements in $\mathrm{Hom}_\Lambda(V,W)$ which are invariant under G is precisely $\mathrm{Hom}_{\Lambda G}(V,W)$.

We may also proceed as in 3.3, 3.7. Let V and W be G-spaces over C which admit structure maps j_V, j_W such that $j_V^2 = \epsilon_V$, $j_W^2 = \epsilon_W$. Then $\mathrm{Hom}_C(V,W)$ admits a structure map j, given by

$$jh = j_W h j_V^{-1}.$$

(Note that $\mathrm{Hom}_C(V,W)$ is covariant in W and contravariant in V.) We have $j^2 = \epsilon_V \epsilon_W$. We can separate three cases.

(i) $\epsilon_V = \epsilon_W = +1$. The Hom of two real representations is real. The construction amounts to taking two G-spaces V,W over R and forming $\mathrm{Hom}_R(V,W)$. In fact, if we form $\mathrm{Hom}_C(cV, cW)$, then the +1 eigenspace of j may be identified with $\mathrm{Hom}_R(V,W)$; thus

$$\mathrm{Hom}_C(cV, cW) \cong c\mathrm{Hom}_R(V,W).$$

(ii) $\epsilon_V = +1$, $\epsilon_W = -1$. The Hom of a real representation into a quaternionic representation is quaternionic. The case $\epsilon_V = -1$, $\epsilon_W = +1$ is similar. We leave it to the reader to interpret the construction in these cases along the lines of 3.7(ii).

(iii) $\epsilon_V = \epsilon_W = -1$. The Hom of two q aternionic representations is real. The construction amounts to taking two G-spaces V,W over Q and forming $\mathrm{Hom}_Q(V,W)$. In fact, if we form $\mathrm{Hom}_C(c'V, c'W)$, then the +1 eigenspace of j is $\mathrm{Hom}_Q(V,W)$; thus

$$\mathrm{Hom}_C(c'V, c'W) \cong c\mathrm{Hom}_Q(V,W).$$

3.10 COROLLARY.

(i) If V and W are two G-spaces over R then

$$\dim_C \mathrm{Hom}_{CG}(cV, cW) = \dim_R \mathrm{Hom}_{RG}(V,W).$$

(ii) If V and W are two G-spaces over Q then

$$\dim_C \mathrm{Hom}_{CG}(c'V, c'W) = \dim_R \mathrm{Hom}_{QG}(V,W).$$

This follows immediately from 3.9(i) and (iii), by looking at the subspaces of elements invariant under G.

3.11 PROPOSITION.

(i) All our Hom's are compatible with the maps c,c'.

(ii) Hom is bilinear over the direct sum $\oplus$.

A particular case of 3.9 is important.

3.12 DEFINITION. Given a G-space V over C, we define its <u>dual</u> V^* by

$$V^* = \mathrm{Hom}_C(V, C).$$

Here the target space C is given the trivial operations from G: $gz = z$ for all $g \in G$ and $z \in C$. The G-space C is real; it follows that the dual of a real representation is real, and the dual of a quaternionic representation is quaternionic.

The general case 3.9 may be reduced to the special case 3.12.

3.13 LEMMA. We have an isomorphism

$$\mathrm{Hom}_C(V, W) \cong V^* \otimes_C W$$

commuting with the action of G and with the structure maps j (if any).

<u>Proof</u>. The isomorphism sends $V^* \otimes W$ into the map h, where

$$h(v) = (v^*v)w.$$

In dealing with compact topological groups, one of our best weapons is integration.

3.14 INTEGRATION. Let G be a compact topological group. Then for each continuous function $f : G \to R$ we can define a real number

$$\int_G f = \int_{g \in G} f(g)$$

so as to satisfy the following conditions.

(i) $\int_G$ has the usual properties of an integral, that is, it is a positive linear functional.

(ii) $\int_G 1 = 1.$

(iii) The integral is invariant under left and right translations; that is, for each $x \in G$ we have

$$\int_{y \in G} f(xy) = \int_{y \in G} f(y)$$

$$\int_{y \in G} f(yx) = \int_{y \in G} f(y).$$

Similarly, we may integrate functions which take values in any finite-dimensional vector space over R so as to obtain values lying in that vector space; and if we do so, integration commutes with linear maps.

If G is a Lie group then the integral is slightly easier to construct than if G is a more general topological group. We will not discuss this here, but refer the reader to [13,14,20].

The first function which asks to be integrated is

$$\theta : G \to \mathrm{Hom}_\Lambda(V, V).$$

3.15 PROPOSITION. Suppose given a representation $\theta : G \to \mathrm{Hom}_\Lambda(V,V)$. Then

$$I = \int_G \theta \in \mathrm{Hom}_\Lambda(V,V)$$

is idempotent ($I^2 = I$) and its image is V_G, the subspace of elements invariant under G.

Proof. For each fixed $v \in V$, the function Hom $(V,V) \to V$ given by $h \to h(v)$ is linear (over R); so it commutes with integretion. That is,

$$Iv = \int_{g \in G} gv.$$

It is now clear that $\mathrm{Im}(I) \subset V_G$; for

$$\begin{aligned} g'(Iv) &= g' \int_{g \in G} gv \\ &= \int_{g \in G} g'gv \quad \text{(since g acts linearly)} \\ &= \int_{g \in G} gv \quad \text{(invariance of integration under left translation)} \\ &= Iv; \end{aligned}$$

so $Iv \in V_G$. Also we have $I|V_G = 1$; for if $v \in V_G$, then

$$\begin{aligned} Iv &= \int_{g \in G} gv \\ &= \int_{g \in G} v \\ &= v. \end{aligned}$$

The proposition follows.

Propositions 3.16 and 3.18 may be viewed as applications of the principle embodied in 3.15; they could equally easily be proved directly.

3.16 PROPOSITION. Let G be a compact topological group and let V be a G-space over C. Then we can give V a positive definite Hermitian form H which is invariant under G. Moreover, if V carries a structure map j, we can choose H so that

$$H(jv,jw) = \overline{H(v,w)}.$$

The reader who wishes to do so may check that if V has a structure map j, then a Hermitian form with the property stated amounts to a Hermitian form over $\Lambda = R$ or Q according to the case. The statement we have given avoids separating cases, and is convenient for later use.

Proof. Consider the space L of Hermitian forms H on V. This is a vector-space over R, and G acts on it by

$$(gH)(v,w) = H(g^{-1}v, g^{-1}w).$$

By 3.15, if we take any Hermitian form H and integrate gH, we get a Hermitian form invariant under G, given by

$$K(v,w) = \int_{g\in G} H(g^{-1}v, g^{-1}w).$$

If we start by choosing H to be positive definite, then K is positive definite.

Now suppose that V has a structure map j, and that we begin by choosing a positive definite Hermitian form H invariant under G. Then we can construct a new form by integrating over the Z_2 or Z_4 group generated by j; the formula is

$$K(v,w) = \frac{1}{2}(H(v,w) + \overline{H(jv,jw)}).$$

This form has the required properties.

If we impose on V an invariant Hermitian form, then we can choose in V an orthonormal basis. Thus we can regard $\theta : G \to \mathrm{Aut}\, V$ as taking values not merely in $GL(n,C)$, but in $U(n)$. We then speak of a unitary representation. Similarly for orthogonal and symplectic representations in the cases $\Lambda = R$ and Q.

3.17 COROLLARY. If G is compact and $\Lambda = C$, then $V^* \cong tV$.

Proof. Impose on V an invariant positive definite Hermitian form H. To be explicit, suppose that $H(v,w)$ is conjugate-linear in v and linear in w. Then we can define

$$\alpha : tV \to V^*$$

by

$$(\alpha v)w = H(v,w),$$

and α is a G-isomorphism over C.

3.18 PROPOSITION. If G is a compact group, then every G-space V is projective. That is, suppose given the following diagram of ΛG-maps, in which β is onto.

$$\begin{array}{ccc} & & X \\ & & \downarrow \beta \\ V & \xrightarrow{\alpha} & Y \end{array}$$

Then there is a ΛG-map $\gamma : V \to X$ such that the following diagram is commutative.

$$\begin{array}{ccc} & \overset{\gamma}{\nearrow} & X \\ & & \downarrow \beta \\ V & \xrightarrow{\alpha} & Y \end{array}$$

Proof. Consider $\mathrm{Hom}_\Lambda (V,X)$, made into a G-space as in 3.9. By 3.15, if we take any Λ-map $\delta : V \to X$ and integrate $g\delta$, we get a Λ-map γ which is invariant under G, that is, a ΛG-map. It is given by

$$\gamma = \int_{g \in G} (\theta_X g)\delta(\theta_V g^{-1}).$$

We can choose δ to be a Λ-map such that $\beta\delta = \alpha$. Then we have

$$\begin{aligned} \beta\gamma &= \beta\int_{g \in G} (\theta_X g)\delta(\theta_V g^{-1}) \\ &= \int_{g \in G} \beta(\theta_X g)\delta(\theta_V g^{-1}) \\ &= \int_{g \in G} (\theta_Y g)\beta\delta(\theta_V g^{-1}) \end{aligned}$$

$$= \int_{g \in G} (\theta_Y g) \alpha (\theta_V g^{-1})$$

$$= \int_{g \in G} \alpha$$

$$= \alpha .$$

3.19 DEFINITION. A non-zero G-space V is <u>reducible</u> if some proper subspace of V is a G-space; otherwise <u>irreducible</u>.

3.20 THEOREM. If G is a compact group, every G-space V is the direct sum of irreducible G-spaces.

<u>Proof</u>. By induction over $\dim_\Lambda V$; so assume the result true for G-spaces W with $\dim_\Lambda W < \dim_\Lambda V$. It will now be sufficient to show that if V is reducible, then it is the direct sum of two subspaces of less dimension. Suppose that V has a proper subspace S which is a G-space; then 3.18 shows that the exact sequence

$$0 \to S \to V \to V/S \to 0$$

splits, so we have a ΛG-isomorphism

$$V \cong S \oplus V/S.$$

Alternatively, if $\Lambda = C$ we may complete the argument by imposing on V a Hermitian form H which is invariant under G, and taking T to be the orthogonal complement of S; then

$$V = S \oplus T.$$

If V has a structure map j, and S is closed under j and H is as in 3.16, then T is closed under j.

3.21 EXAMPLE. We will show that 3.20 does not hold for groups which are not compact.

Embed R^1 in R^2 as the subspace of vectors $\begin{bmatrix} x \\ 0 \end{bmatrix}$. Let G be the subgroup of GL(2,R) which stabilises R^1. Equivalently, G is the set of matrices $\begin{bmatrix} a & b \\ 0 & c \end{bmatrix}$ with $ac \neq 0$.
Then R^2 is a reducible G-space. However, no other proper subspace of R^2 is stable under G, so R^2 does not split as the direct sum of irreducible G-spaces.

Alternatively, to get a "minimal" counter-example, take the group G to be the set of matrices $\begin{bmatrix} 1 & b \\ 0 & 1 \end{bmatrix}$.

Next we shall need to know to what extent the decomposition of a G-space into irreducible summands is unique (3.24). For this purpose we need the following classical result.

3.22 (SCHUR'S LEMMA). Let G be any topological group.

(i) If $f : V \to W$ is a ΛG-map and V, W are irreducible then f is either zero or an isomorphism.

(ii) If $\Lambda = C$, $f : V \to V$ is a CG-map and V is irreducible then $fv = \lambda v$ for some constant $\lambda \in C$.

(In the second case we may write $f = \lambda$.)

Proof.

(i) Since V and W are irreducible, Ker f is V or 0 and Im f is 0 or W. The result follows.

(ii) Consider $f - \lambda : V \to V$, where λ runs through C. This map is singular for some λ. By (i), $f - \lambda$ is then zero. Thus $f = \lambda$.

3.23 COROLLARY. Let V and W be irreducible ΛG-spaces.

(i) If V and W are inequivalent then $\mathrm{Hom}_{\Lambda G}(V, W) = 0$.

(ii) If V and W are equivalent and $\Lambda = C$, then $\dim_C \mathrm{Hom}_{CG}(V, W) = 1$.

(iii) If V and W are equivalent and $\Lambda = R$ or Q, then $\dim_R \mathrm{Hom}_{\Lambda G}(V, W) \geq 1$.

Proof. For (iii) we observe that $\mathrm{Hom}_{\Lambda G}(V, W)$ contains at least one isomorphism.

For the next proposition, let G be any topological group, and let V_i run over the inequivalent irreducible ΛG-spaces (as i runs over some set of indices I). Let m_i, n_i be non-negative integers, of which all but a finite number are zero. Let $m_i V_i$ be the direct sum of m_i copies of V_i, and similarly for $n_i V_i$.

3.24 THEOREM. If $\bigoplus_i m_i V_i$ is equivalent to $\bigoplus_i n_i V_i$, then $m_i = n_i$ for all i.

Proof. Suppose

$$\bigoplus_i m_i V_i = \bigoplus_i n_i V_i .$$

Then

$$\mathrm{Hom}_{\Lambda G}(V_j, \bigoplus_i m_i V_i) \cong \mathrm{Hom}_{\Lambda G}(V_j, \bigoplus_i n_i V_i),$$

that is,

$$\bigoplus_i m_i \mathrm{Hom}_{\Lambda G}(V_j, V_i) \cong \bigoplus_i n_i \mathrm{Hom}_{\Lambda G}(V_j, V_i).$$

Using 3.23(i), we get

$$m_j \mathrm{Hom}_{\Lambda G}(V_j, V_j) \cong n_j \mathrm{Hom}_{\Lambda G}(V_j, V_j).$$

Taking the dimension of both sides and using 3.23(ii) or (iii), we get $m_j = n_j$.

If G is compact and $\Lambda = C$ we can express the situation which arises here in the following way (which we need for later use). For any G-space V over C we can form

$$\bigoplus_i \mathrm{Hom}_{CG}(V_i, V) \otimes_C V_i .$$

This is a finite sum, since $\mathrm{Hom}_{CG}(V_i, V)$ is zero for all but a finite number of i, by 3.20 and 3.23(i). We can define

$$\mu : \bigoplus_i \mathrm{Hom}_{CG}(V_i, V) \otimes_C V_i \to V$$

by evaluation:

$$\mu(h_i \otimes v_i) = h_i(v_i).$$

We make G act on $\bigoplus_i \mathrm{Hom}_{CG}(V_i, V) \otimes_C V_i$ by

$$g(h_i \otimes v_i) = h_i \otimes gv_i.$$

Then μ is a G-map over C.

3.25 LEMMA. Assume G compact and $\Lambda = C$. Then the map

$$\mu : \bigoplus_i \mathrm{Hom}_{CG}(V_i, V) \otimes_C V_i \to V$$

is an isomorphism.

Proof. If V is irreducible the result is immediate by 3.23. Pass to direct sums and use 3.20.

3.26 DEFINITION. Let G be a compact topological group. Then $K_\Lambda(G)$ is the free abelian group generated by the equivalence classes of irreducible G-spaces over Λ.

Tnus an element of $K_\Lambda(G)$ is a formal linear combination $\sum_i n_i V_i$, in which the V_i are the equivalence classes of irreducible G-spaces over Λ, and the n_i are integers (positive, negative or zero) which are zero for all but a finite number of i.

By 3.20 and 3.24, the equivalence classes of G-spaces over Λ are in 1-1 correspondence with those elements $\Sigma n_i V_i$ in $K_\Lambda(G)$ such that $n_i \geq 0$ for all i.

An element of $K_\Lambda(G)$ is called a <u>virtual representation</u> or <u>virtual G-space</u>.

The operations c, c', r, q and t of 3.5 induce homomorphisms of abelian groups as displayed in the following diagram, which is not commutative.

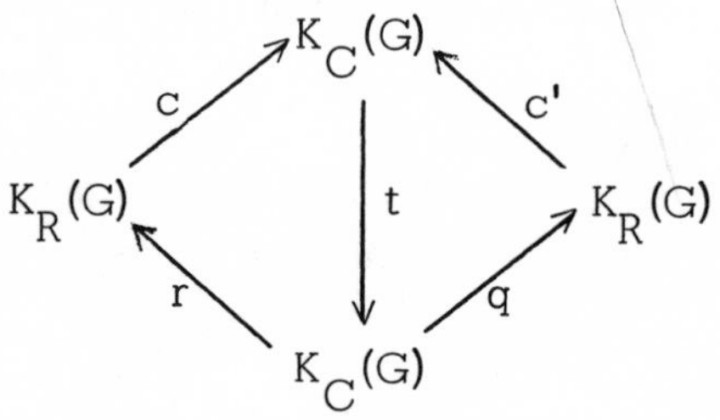

The equations of 3.6 continue to hold.

3.27 PROPOSITION. The maps

$$c : K_R(G) \to K_C(G)$$

$$c' : K_Q(G) \to K_C(G)$$

are mono.

<u>Proof</u>. $rc = 2$, $qc' = 2$ and $K_R(G)$, $K_Q(G)$ are free abelian.

We shall normally regard $K_R(G)$ and $K_Q(G)$ as embedded in $K_C(G)$ by c and c'.

3.28 COROLLARY

(i) If V and W are two G-spaces over R such that $cV \cong cW$, then $V \cong W$.

(ii) If V and W are two G-spaces over Q such that $c'V \cong c'W$, then $V \cong W$.

This follows immediately from 3.27, but in case it seems to spring from nowhere we also give a direct proof. Suppose given two G-spaces V, W over C, which admit structure maps j_V, j_W such that $j_V^2 = j_W^2$. We suppose given a CG-isomorphism $f : V \to W$ which does not necessarily commute with j, and we wish to construct a CG-isomorphism which does commute with j. We can construct CG-maps which do commute with j by starting from f, or if, and integrating over the Z_2 or Z_4 group generated by j. The formulae are

$$f' = \frac{1}{2}(f + j_W f j_V^{-1})$$

$$f'' = \frac{1}{2} i (f - j_W f j_V^{-1}).$$

We have $f' - if'' = f$. So $\det(f' + zf'')$ is a polynomial in z which is not identically zero (for it is non-zero for $z = -i$). Therefore there is some real x for which $\det(f' + xf'') \neq 0$. Then $f' + xf''$ is a CG-isomorphism which commutes with j.

TRIVIAL EXERCISE. If V admits a structure map j, then it also admits -j, and clearly forgetting j gives the same result as forgetting -j. Display a CG-automorphism sending j into -j.

If $\Lambda = C$, we can make $K_C(G)$ into a ring by using the tensor product of G-spaces over C; we then call it the <u>representation ring</u> of G. If x lies in $K_\Lambda(G) \subset K_C(G)$, where $\Lambda = R$ or Q, and y lies in $K_{\Lambda'}(G) \subset K_C(G)$, where $\Lambda' = R$ or Q, then the product xy behaves as described in 3.7.

The standard method of studying $K_C(G)$, and indeed the standard method of proving 3.24, is the study of <u>characters</u>. To define these, we need the <u>trace</u>.

3.29 DEFINITION. Let V be a finite-dimensional vector space over C, and let $f : V \to V$ be a linear map. Then we may define Tr f, the <u>trace</u> of f, in two ways.

(i) Take a base of V, so that f corresponds to a matrix M_{ij}. Set $\mathrm{Tr}\, f = \sum_i M_{ii}$. This is invariant under change of base, since

$$\sum_{i,j,k} T_{ij} M_{jk} (T^{-1})_{ki} = \sum_{j,k} I_{jk} M_{jk} = \sum_j M_{jj}.$$

(ii) (Bourbaki) We have an isomorphism

$$\alpha : V^* \otimes V \to \mathrm{Hom}_C(V, V)$$

given by $(\alpha(v^* \otimes w))v = (v^*v)w$, as in 3.13.

We have an evaluation map

$$\epsilon : V^* \otimes V \to C$$

given by $\epsilon(v^* \otimes w) = v^*w$. Define $\mathrm{Tr}\, f = \epsilon\alpha^{-1} f$.

It is easy to check that the two definitions are equivalent. The principal properties of the trace are as follows.

3.30 PROPOSITION.

(i) $\mathrm{Tr} : \mathrm{Hom}_C(V,V) \to C$ is a linear map.

(ii) Consider $V \xrightarrow{\beta} W \xrightarrow{\gamma} V$. Then $\mathrm{Tr}(\beta\gamma) = \mathrm{Tr}(\gamma\beta)$.

(iii) Consider $\beta \oplus \gamma : V \oplus W \to V \oplus W$.

Then $\mathrm{Tr}(\beta \oplus \gamma) = \mathrm{Tr}\beta + \mathrm{Tr}\gamma$.

(iv) Consider $\beta \otimes \gamma : V \otimes W \to V \otimes W$.

Then $\mathrm{Tr}(\beta \otimes \gamma) = \mathrm{Tr}\beta \cdot \mathrm{Tr}\gamma$.

(v) Given $\beta : V \to V$, define $\beta^* : V^* \to V^*$ by $(\beta^* v^*)v = v^*(\beta v)$, as usual. Then $\mathrm{Tr}\beta^* = \mathrm{Tr}\beta$.

(vi) Given $\beta : V \to V$, let $t\beta : tV \to tV$ be as in 3.5(v). Then $\mathrm{Tr}(t\beta) = \overline{\mathrm{Tr}\beta}$.

(vii) If $\beta : V \to V$ is idempotent, then

$$\mathrm{Tr}\beta = \dim_C \mathrm{Im}\beta.$$

The proof may safely be left to the reader.

3.31 DEFINITION. Given a G-space over C, we define its <u>character</u> $\chi_V : G \to C$ by

$$\chi_V(g) = \mathrm{Tr}\,\theta g.$$

It is clear that χ_V depends only on the equivalence class of V.

If V is a G-space over R or Q, we define its character to be that of the complex G-space cV or c'V as the case may be. (In the case $\Lambda = R$ it would be equivalent to consider the trace over R, but this doesn't work so well for $\Lambda = Q$.)

3.32 PROPOSITION

(i) $\chi_V : G \to C$ is continuous.

(ii) $\chi_V(xyx^{-1}) = \chi_V(y)$.

(iii) $\chi_{V\oplus W}(g) = \chi_V(g) + \chi_W(g)$.

(iv) $\chi_{V\otimes W}(g) = \chi_V(g) \cdot \chi_W(g)$.

(v) $\chi_{V^*}(g) = \chi_V(g^{-1})$.

(vi) $\chi_{tV}(g) = \overline{\chi_V(g)}$; if V is real or quaternionic then $\overline{\chi_V(g)} = \chi_V(g)$.

(vii) $\chi_V(e) = \dim_C(V)$.

Each part follows from the corresponding part of 3.31; the second half of (vi) uses also the equations tc = c, tc' = c' from 3.6.

3.33 PROPOSITION. Assume G compact.

(i) $\chi_V(g^{-1}) = \chi_{V*}(g) = \chi_{tV}(g) = \overline{\chi_V(g)}$.

(ii) $\int_{g \in G} \chi_V(g) = \dim_C V_G$, where V_G is the subspace of elements of V invariant under G.

Proof.

(i) See 3.17.

(ii) Since Tr is linear, we have

$$\int_{g\in G} \mathrm{Tr}\,\theta g = \mathrm{Tr} \int_{g \in G} \theta g$$

$$= \mathrm{Tr}\ I \qquad \text{(see 3.15)}$$

$$= \dim_C \mathrm{Im}\ I \qquad \text{(see 3.30(vii))}$$

$$= \dim_C V_G \qquad \text{(see 3.15).}$$

3.34 THEOREM. (Orthogonality relations for characters.)

(i) Let G be compact and let V, W be G-spaces over Λ.

Then

$$\int_{g\in G} \overline{\chi_V(g)}\chi_W(g) = \dim \mathrm{Hom}_{\Lambda G}(V,W)$$

$$= d \text{ say,}$$

where the dimension is taken over C if $\Lambda = C$, over R if $\Lambda = R$ or Q.

(ii) Now assume that V and W are irreducible. If V and W are inequivalent we have $d = 0$. If V and W are equivalent and $\Lambda = C$ we have $d = 1$. If V and W are equivalent and $\Lambda = R$ or Q we have $d \geq 1$.

Proof.

(i) By 3.10 the cases $\Lambda = R$ and Q follow immediately from the case $\Lambda = C$. So suppose $\Lambda = C$, and consider $H = \mathrm{Hom}_C(V,W)$. We have

$$\begin{aligned}
\dim_C \mathrm{Hom}_{CG}(V,W) &= \dim_C H_G \\
&= \int_{g\in G} \chi_H(g) && \text{by 3.33(ii)} \\
&= \int_{g\in G} \chi_{V^*\otimes W}(g) && \text{by 3.13} \\
&= \int_{g\in G} \overline{\chi_V(g)}\, \chi_W(g) && \text{by 3.32(iv), 3.33(i).}
\end{aligned}$$

(ii) See 3.23.

Let us choose one irreducible ΛG-space V_i in each equivalence class, as in 3.24, and let χ_i be its character. Then the functions χ_i are orthogonal, and therefore:

3.35 COROLLARY. The functions χ_i are linearly independent.

This fact can evidently be used to give a second proof

of 3.24. If

$$\bigoplus_i m_i V_i \cong \bigoplus_i n_i V_i,$$

then their characters are equal, so

$$\sum_i m_i \chi_i = \sum_i n_i \chi_i,$$

and $m_i = n_i$ for each i. But by using 3.34(i), we see that this proof coincides with the first proof.

Let C(G) be the set of continuous functions $f : G \to C$.

3.36 DEFINITION. Such an f is called a <u>class function</u> if $f(xyx^{-1}) = f(y)$.

We write Cl(G) for the set of class functions. We make Cl(G) into a ring by pointwise addition and multiplication of functions.

Characters are class functions, by 3.32(i) and (ii). We can define a homomorphism of abelian groups

$$\chi : K_C(G) \to Cl(G)$$

by

$$\chi(\sum_i n_i V_i) = \sum_i n_i \chi_i .$$

For every G-space V we have

$$\chi(V) = \chi_V,$$

by 3.32(iii). χ is a homomorphism of rings by 3.32(iv).

3.37 PROPOSITION. $\chi : K_C(G) \to Cl(G)$ is a monomorphism.

Proof. See 3.35.

The image of χ is called the character ring of G. It is natural to ask how large a part of Cl(G) it is; and we will see that it is as large as could be hoped (3.47). For this purpose we need the Peter-Weyl theorem.

We recall that classically the Peter-Weyl theorem is stated in terms of component-functions $M_{ij}(g)$ of matrix representations $M(g)$. Clearly such a function is obtained by taking a matrix representation $M : G \to GL(n, C)$ and composing with a linear map $GL(n, C) \to C$, namely projection onto the (i, j)th component. We therefore introduce the following lemma.

3.38 LEMMA. The vector-space dual to $\mathrm{Hom}_C(V, W)$ is $\mathrm{Hom}_C(W, V)$, where the pairing between $\alpha \in \mathrm{Hom}_C(V, W)$ and $\beta \in \mathrm{Hom}_C(W, V)$ is given by

$$\langle \beta, \alpha \rangle = \mathrm{Tr}(\alpha\beta) = \mathrm{Tr}(\beta\alpha).$$

Proof. We have $\mathrm{Hom}_C(V, W) \cong V^* \otimes W$, and therefore its dual space is

$$V \otimes W^* \cong W^* \otimes V \cong \mathrm{Hom}_C(W, V).$$

To check that the pairing is as claimed is precisely what the student should already have done in proving $Tr(\alpha\beta) = Tr(\beta\alpha)$ (3.30(ii)).

3.39 THEOREM (F. Peter and H. Weyl) [15]. Let G be a compact topological group. Then every continuous function $f : G \to C$ can be uniformly approximated by functions of the form $Tr(\alpha\theta(g))$, where θ runs over representations $\theta : G \to Hom_C(V,V)$ and α runs over $Hom_C(V,V)$.

The proof will occupy 3.40 to 3.44. Actually our line of proof will approximate f by functions of the form $Tr(\alpha\theta(g^{-1}))$; but this makes no difference, since we can begin by replacing f with f', where $f'(g) = f(g^{-1})$.

The proof is based on the following ideas. We make G act on C(G) by

$$(gf)(x) = f(g^{-1}x) .$$

Then C(G) is an infinite-dimensional representation of G. However, we can find certain finite-dimensional subspaces of C(G) stable under G, by using the theory of integral operators

$$\int_{y \in G} k(x,y) f(y) .$$

We make G act on $C(G \times G)$ by

$$(gk)(x,y) = k(g^{-1}x, g^{-1}y).$$

If the "kernel" k is invariant under G, then the integral operator gives a G-map from C(G) to C(G), and hence its eigenspaces are stable under G. In the case at issue they are finite-dimensional (3.42) and this provides the necessary representations.

We now start work.

3.40 LEMMA. Let G be a compact group and $f \in C(G)$. Then f can be uniformly approximated by functions of the form

$$v(x) = \int_{y \in G} k(x,y) f(y)$$

with k real symmetric and invariant under G.

Proof. There is a neighbourhood U of e in G such that

$$|f(x) - f(y)| \leq \epsilon \qquad \text{for } x^{-1}y \in U$$

and $U^{-1} = U$. Let $\mu : G \to R$ be a continuous function such that

$$\mu(x) = 0 \qquad \text{for } x \notin \dot{U},$$

$$\mu(x) \geq 0,$$

$$\mu(x^{-1}) = \mu(x) \qquad \text{and}$$

$$\int_{x \in G} \mu(x) = 1.$$

Let

$$k(x,y) = \mu(x^{-1}y).$$

Then k is real, symmetric and invariant under G. Also

$$|\mu(x^{-1}y)f(x) - \mu(x^{-1}y)f(y)| \leq \epsilon\mu(x^{-1}y)$$

everywhere. Integrating over $y \in G$, we get

$$|f(x) - v(x)| \leq \epsilon$$

where

$$v(x) = \int_{y\in G} k(x,y)f(y).$$

It will now be sufficient to approximate such functions $v(x)$.

3.41 THEOREM. Assume that k is Hermitian and $u \in C(G)$. Then the function

$$v(x) = \int_{y\in G} k(x,y)u(y)$$

can be uniformly approximated by a finite linear combination of eigenfunctions of k corresponding to non-zero eigenvalues.

The eigenfunctions corresponding to the eigenvalue λ are, of course, the functions w such that

$$\int_{y\in G} k(x,y)w(y) = \lambda w(x).$$

Proof. See [16, p. 117, 127]. (Smithies considers integral equations on [a,b], but the results are unchanged for integral equations on a compact manifold.)

Note also that even if we were to consider some class

of functions larger than C(G), for example $L^2(G)$, the eigenfunctions would be continuous, since k is continuous.

3.42 THEOREM. Assume that k is Hermitian and $\lambda \neq 0$. Then the vector space V of eigenfunctions corresponding to λ has finite dimension. Indeed the sum $\sum_i |\lambda_i|^2$, in which $|\lambda|^2$ is repeated with appropriate multiplicity, is convergent.

Proof. See [16, pp. 48,102,112].

3.43 LEMMA. In 3.42, assume further that k is invariant under G. Then every element of V can be written in the required form $\mathrm{Tr}(\alpha\theta(g^{-1}))$.

Proof. The space V is a finite-dimensional G-space. Let $v \in V$. Define a linear map

$$\beta : \mathrm{Hom}_C(V,V) \to C$$

as follows: if $h \in \mathrm{Hom}_C(V,V)$, then

$$\beta(h) = (hv)(e).$$

Then we have

$$\beta(\theta(g^{-1})) = v(g).$$

By 3.38, the element β corresponds to an element $\alpha \in \mathrm{Hom}_C(V,V)$ such that

$$\mathrm{Tr}(\alpha\theta(g^{-1})) = v(g).$$

3.44 LEMMA. The set of continuous function $G \to C$ which can be written in the form $\mathrm{Tr}(\alpha\theta(g^{-1}))$ is closed under linear combinations.

Proof. Suppose given

$\theta' : G \to \mathrm{Hom}_C(V', V')$, $\quad \theta'' : G \to \mathrm{Hom}_C(V'', V'')$

$\alpha' \in \mathrm{Hom}_C(V', V')$, $\quad \alpha'' \in \mathrm{Hom}_C(V'', V'')$

and $\lambda', \lambda'' \in C$. Form $V = V' \oplus V''$ and consider $\alpha = \lambda'\alpha' \oplus \lambda''\alpha'' \in \mathrm{Hom}_C(V, V)$. Then we have

$$\mathrm{Tr}(\alpha\theta(g^{-1})) = \lambda'\,\mathrm{Tr}(\alpha'\theta'(g^{-1})) + \lambda''\,\mathrm{Tr}(\alpha''\theta''(g^{-1})).$$

This completes the proof of 3.39; any function $f(x)$ can be uniformly approximated by a function $v(x)$ as in 3.40, which in turn can be uniformly approximated by a linear combination of eigenfunctions by 3.41; and this can be written in the required form $\mathrm{Tr}(\alpha\theta(g^{-1}))$ by 3.42-3.44.

3.45 REMARK. If $\alpha : V \to V$ is a G-map, then $\mathrm{Tr}(\alpha\theta(g))$ is a class function.

Proof. $\mathrm{Tr}(\alpha\theta(xyx^{-1})) = \mathrm{Tr}(\alpha(\theta x)(\theta y)(\theta x^{-1}))$

$$= \mathrm{Tr}((\theta x^{-1})\alpha(\theta x)(\theta y)) \qquad (3.30\text{(ii)})$$

$$= \mathrm{Tr}(\alpha\theta(y))$$

since α is a G-map.

There is a converse to this remark.

3.46 PROPOSITION. Let G be a compact group. Then every class function $f : G \to C$ can be uniformly approximated by functions of the form $\mathrm{Tr}(\beta\theta(g))$, where θ runs over representations $\theta : G \to \mathrm{Hom}_C(V,V)$ and β runs over $\mathrm{Hom}_{CG}(V,V)$, that is, β runs over G-maps.

Proof. By 3.39 we can find $\theta : G \to \mathrm{Hom}_C(V,V)$ and $\alpha \in \mathrm{Hom}_C(V,V)$ such that

$$|f(x) - \mathrm{Tr}(\alpha\theta(x))| \le \epsilon .$$

If f is a class function we can substitute $y^{-1}xy$ for x and get

$$|f(x) - \mathrm{Tr}(\alpha\theta(y^{-1}xy))| \le \epsilon .$$

Arguing as in 3.45, this gives

$$|f(x) - \mathrm{Tr}((\theta y)\alpha(\theta y^{-1})(\theta x))| \le \epsilon .$$

Integrate over y; we find

$$|f(x) - \mathrm{Tr}(\beta(\theta x))| \le \epsilon$$

where

$$\beta = \int_{y \in G} (\theta y)\alpha(\theta y^{-1}) .$$

But as in 3.18, β is a G-map.

3.47 THEOREM. Let G be a compact topological group. Then every class function $f : G \to C$ can be uniformly approximated by a linear combination $\sum_i \lambda_i \chi_i$ of irreducible complex characters.

Proof. Let θ and β be as in 3.46, and let V_i and μ be as in 3.25. Then the G-map $\beta : V \to V$ induces (say)

$$\beta_i : \mathrm{Hom}_{CG}(V_i, V) \to \mathrm{Hom}_{CG}(V_i, V).$$

We have the following commutative diagram.

$$\begin{array}{ccc} \bigoplus_i \mathrm{Hom}_{CG}(V_i, V) \otimes V_i & \xrightarrow[\cong]{\mu} & V \\ \Big\downarrow \bigoplus_i \beta_i \otimes \theta_i(g) & & \Big\downarrow \beta\theta(g) \\ \bigoplus_i \mathrm{Hom}_{CG}(V_i, V) \otimes V_i & \xrightarrow[\cong]{\mu} & V \end{array}$$

Therefore

$$\mathrm{Tr}(\beta\theta(g)) = \sum_i (\mathrm{Tr}\beta_i) \cdot \mathrm{Tr}(\theta_i g),$$

which has the required form $\sum_i \lambda_i \chi_i(g)$.

It is natural to ask for the analogue of 3.47 over R or Q. By 3.6 we have $tcV \cong cV$, $tc'V \cong c'V$; so by 3.32(vi) the character of a representation over R or Q is real, and by 3.33(i) it satisfies $\chi(g^{-1}) = \chi(g)$.

3.48 COROLLARY. Every class function $f : G \to R$ such that $f(g) = f(g^{-1})$ can be uniformly approximated by an R-linear combination of characters of representations over R, or by an R-linear combination of characters of representations over Q.

Proof. Let $f : G \to R$ be a class function such that $f(g) = f(g^{-1})$. By 3.47 we can find complex λ_i such that

$$|f(g) - \sum_i \lambda_i \chi_i(g)| < \epsilon.$$

Since $f(g) = f(g^{-1})$ we have

$$|f(g) - \sum_i \lambda_i \chi_i(g^{-1})| < \epsilon,$$

or using 3.33(i)

$$|f(g) - \sum_i \lambda_i \overline{\chi_i(g)}| < \epsilon.$$

Since f is real we also have

$$|f(g) - \sum_i \bar{\lambda}_i \overline{\chi_i(g)}| < \epsilon$$

$$|f(g) - \sum_i \bar{\lambda}_i \chi_i(g)| < \epsilon.$$

Therefore

$$|f(g) - \sum_i \frac{1}{4}(\lambda_i + \bar{\lambda}_i)(\chi_i(g) + \overline{\chi_i(g)})| < \epsilon.$$

But here $\frac{1}{4}(\lambda_i + \bar{\lambda}_i)$ is a real coefficient and $\chi_i + \bar{\chi}_i$ is the character of the real representation rV_i or of the quaternionic representation qV_i (see 3.6).

3.49 COROLLARY. Every class function $f : G \to C$ such that $f(g) = f(g^{-1})$ can be uniformly approximated by a C-linear combination of characters of representations over R or by a

C-linear combination of representations over Q.

<u>Proof</u>. Approximate the real and imaginary parts of f by 3.48.

We now consider in greater detail which complex representations are real or quaternionic.

3.50 THEOREM. A representation V over C is real if and only if there exists a non-singular symmetric bilinear form $B : V \otimes V \to C$ which is invariant under G.

A representation V over C is quaternionic if and only if there exists a non-singular skew-symmetric bilinear form $B : V \otimes V \to C$ which is invariant under G.

<u>Proof</u>. First suppose that V carries a structure map j such that $j^2 = \epsilon = \pm 1$. By 3.16 we can impose on V a positive definite Hermitian form H which is invariant under G and satisfies

$$H(jv, jw) = \overline{H(v,w)}.$$

Define

$$B(v,w) = H(jv,w).$$

Then B is clearly bilinear, non-singular and invariant under G. Also we have

$$\begin{aligned} B(w,v) &= H(jw,v) \\ &= \overline{H(v,jw)} \\ &= H(jv,j^2w) \\ &= \epsilon H(jv,w) \\ &= \epsilon B(v,w). \end{aligned}$$

So B is symmetric or antisymmetric according to the sign of ϵ.

We now seek to reverse this argument. Suppose given on V a non-singular bilinear form $B : V \otimes V \to C$ which is invariant under G and satisfies

$$B(w,v) = \epsilon B(v,w),$$

where $\epsilon = \pm 1$. By 3.16, we can also suppose that V carries a positive definite Hermitian form H which is invariant under G. Then we can define $f : V \to V$ by

$$B(v,w) = H(fv,w).$$

The map f is conjugate-linear, a G-map and a 1-1 correspondence. The property of B gives

$$\begin{aligned} H(fv,w) &= B(v,w) \\ &= \epsilon B(w,v) \\ &= \epsilon H(fw,v) \\ &= \epsilon \overline{H(v,fw)}. \end{aligned}$$

Thus

3.51 $H(fv,w) = \epsilon\overline{H(v,fw)}$.

We now define another positive-definite Hermitian form on V by

3.52 $K(v,w) = \overline{H(fv,fw)}$.

The form K is invariant under G. Substituting fv and fw into 3.51, we get

$$H(f^2v,fw) = \epsilon\overline{H(fv,f^2w)},$$

and taking complex conjugates, we get

3.53 $K(fv,w) = \epsilon\overline{K(v,fw)}$.

The space V now splits as the direct sum of eigenspaces V_i for the pair of forms H,K. The eigenvalues are positive real numbers λ_i; for each such λ_i, V_i is the set of v_i such that

3.54 $K(v_i,w) = \lambda_i H(v_i,w)$ for all $w \in V$.

The eigenspaces V_i are stable under G. I claim they are also preserved by f; for we have

$$\begin{aligned} K(fv_i,w) &= \epsilon\overline{K(v_i,fw)} && (3.53) \\ &= \epsilon\lambda_i\overline{H(v_i,fw)} && (3.54) \\ &= \lambda_i H(fv_i,w) && (3.51). \end{aligned}$$

Thus $fv_i \in V_i$.

We also have

$$H(f^2 v_i, w) = \epsilon\overline{H(fv_i, fw)} \qquad (3.51)$$

$$= \epsilon K(v_i, w) \qquad (3.52)$$

$$= \epsilon\lambda_i H(v_i, w) \qquad (3.54).$$

So $f^2 \mid V_i = \epsilon\lambda_i$, where λ_i is real and positive.

Let us now define a map $j : V \to V$ by

$$j \mid V_i = (\lambda_i)^{-\frac{1}{2}} f \mid V_i.$$

Then j is conjugate-linear, a G-map and satisfies $j^2 = \epsilon$. Thus V is real or quaternionic according to the sign of ϵ. This completes the proof.

To sum up, the advantage of structure maps is that they come normalised by the condition $j^2 = \pm 1$; the disadvantage of bilinear maps is that they can be denormalised by a scalar factor for each summand of V.

3.55 DEFINITION. We say that a representation V of G is self-conjugate if $tV = V$. Evidently representations over R and Q are self-conjugate (either using 3.6 or using the fact that the structure map j gives an isomorphism from tV to V).

3.56 PROPOSITION. If a complex irreducible representation V of G is self-conjugate, then it is either real or quaternionic, but not both.

Proof. Consider $V^* \otimes V^*$, the space of bilinear maps from $V \otimes V$ to C. It has an automorphism τ defined by

$$\tau(v^* \otimes w^*) = w^* \otimes v^*.$$

We have $\tau^2 = 1$. So $V^* \otimes V^*$ splits as the direct sum of the +1 and -1 eigenspaces of τ. The +1 eigenspace is the space S^* of symmetric bilinear maps; the -1 eigenspace is the space A^* of antisymmetric bilinear maps.

Now we also have

$$V^* \otimes V^* \cong \operatorname{Hom}_C(V, V^*).$$

By 3.17 we have $V^* \cong tV$, and if V is self-conjugate we have $V^* \cong V$. If V is irreducible then so is V^*, and we have

$$\dim_C \operatorname{Hom}_{CG}(V, V^*) = 1$$

by 3.23. That is, for the elements invariant under G we have

$$\dim_C S^*_G + \dim_C A^*_G = 1.$$

Moreover, a non-zero bllinear map B which is invariant under G corresponds to a non-zero G-map $V \to V^*$, which must be iso; so such a B is non-singular. We conclude that only two cases are possible.

(i) $\dim_C S^*_G = 1$, $\dim_C A^*_G = 0$. In this case V admits a non-singular symmetric bilinear form invariant under G, but not an antisymmetric one.

(ii) $\dim_C S^*_G = 0$, $\dim_C A^*_G = 1$. In this case V admits a non-singular antisymmetric bilinear form invariant under G, but not a symnetric one.

The result follows by 3.50.

3.57 THEOREM. Suppose given a compact group G. Then it is possible to choose representations U_m over R, V_n over C and W_p over Q to satisfy the following conditions.

(i) The inequivalent irreducible representations over R are precisely the U_m, rV_n and $rc'W_p$.

(ii) The inequivalent irreducible representations over C are precisely the cU_m, V_n, tV_n and $c'W_p$.

(iii) The inequivalent irreducible representations over Q are precisely the qcU_m, qV_n and W_p.

Proof. We begin by taking the irreducible complex representations V. First, we can classify them into those such that $tV \cong V$ and those such that $tV \ncong V$. The latter occur in pairs (V, tV), and we choose one V_n out of each pair. The former are either real or quaternionic by 3.56; we choose U_m over R, W_p over Q so that the cU_m and $c'W_p$ give such V. It is clear that this choice makes 3.57(ii) true.

It is also claimed that the representations U_m, rV_n and $rc'W_p$ over R are irreducible, and similarly over Q. In fact, U_m is irreducible because cU_m is so. We have

$$crV_n \cong (1 + t)V_n$$

$$crc'W_p \cong 2c'W_p$$

and neither can be split into real representations because V_n and tV_n are not self-conjugate and $c'W_p$ is not real. Similarly over Q.

It remains only to prove that there can be no further irreducible representations over R or Q. For this purpose we introduce:

3.58 LEMMA. If V and W are inequivalent irreducible representations over R, then no complex irreducible representation can occur as a summand both in cV and in cW. Similarly for c'V and c'W if V and W are over Q.

Proof. $\dim_C \mathrm{Hom}_{CG}(cV, cW) = \dim_R \mathrm{Hom}_{RG}(V, W)$ (3.10)

$= 0$ (3.23).

To complete the proof of 3.57, it remains only to remark that all the complex irreducible representations occur as summands in

$$cU_m$$

$$crV_n = (1 + t)V_n$$

$$crc'W_p = 2c'W_p .$$

Therefore there can be no more irreducible representations over R. Similarly over Q.

There is a classical criterion for deciding whether a complex irreducible representation is real or quaternionic. For this purpose we introduce the following considerations.

3.59 DEFINITION. Let $V^n = V \otimes V \otimes \ldots \otimes V$ (n factors). Let $\lambda^n V$ be the summand of V^n on which the permutation group Σ_n acts by

$$\rho w = (\epsilon \rho) w ,$$

where $\epsilon\rho$ is the sign of ρ; that is, $\lambda^n V$ is the space of anti-symmetric or alternating tensors. The G-space $\lambda^n V$ is called the <u>nth exterior power</u> of V.

Consider the power-sum

$$x_1^k + x_2^k + \ldots + x_m^k$$

in $m \geq k$ variables. This can be written as a polynomial $P_k(\sigma_1, \sigma_2, \ldots, \sigma_k)$ in the elementary symmetric functions σ_i of $x_1, x_2, \ldots, x_m$; the polynomial is actually independent of m, and the formula is valid even for $m < k$.

3.60 DEFINITION. If V is a complex representation of G, we define a virtual representation by

$$\psi^k(V) = P_k(\lambda^1 V, \lambda^2 V, \ldots, \lambda^k V).$$

The polynomial is evaluated in the ring $K_C(G)$.

3.61 LEMMA. If $W = \psi^k V$, then

$$\chi_W(g) = \chi_V(g^k).$$

Of course it is clear that $\chi_V(g^k)$ is a class function; after 3.47 it is natural to ask what it is the character of.

<u>Proof</u>. Impose on V a positive-definite Hermitian form H. Fix g. Then $\theta_V(g)$ is a unitary map, and we can find in V a base of eigenvectors v_i with eigenvalues λ_i. Then V^n admits a base of eigenvectors $v_{i_1} \otimes v_{i_2} \otimes \ldots \otimes v_{i_n}$ with eigenvalues $\lambda_{i_1} \lambda_{i_2} \ldots \lambda_{i_n}$, and similarly for $\lambda^n V$. Hence g acts on $\lambda^n V$ with trace σ_n, the nth elementary symnetric function of the λ_i. Hence

$$\begin{aligned} \chi_W(g) &= P_k(\sigma_1, \sigma_2, \ldots, \sigma_k) \\ &= \lambda_1^k + \lambda_2^k + \ldots \qquad \text{(by definition of } P_k) \\ &= \operatorname{Tr}((\theta_V g)^k) \\ &= \chi_V(g^k). \end{aligned}$$

3.6 THEOREM. Let V be a complex irreducible representation of a compact group G. Then

$$\int_{g \in G} \chi_V(g^2) = \begin{cases} 1 \text{ if V is real} \\ 0 \text{ if V is not self-conjugate} \\ -1 \text{ if V is quaternionic} \end{cases}$$

Proof. In 3.56, instead of considering $V^* \otimes V^* \cong S^* \oplus A^*$, it is equivalent to consider $V \otimes V \cong S \oplus A$. We have

$$P_2(\sigma_1, \sigma_2) = (\sigma_1)^2 - 2\sigma_2$$

and so

$$\begin{aligned} \psi^2(V) &= V^2 - 2A \\ &= S - A. \end{aligned}$$

Therefore

$$\begin{aligned} \int_{g \in G} \chi_V(g^2) &= \int_{g \in G} \chi_S(g) - \chi_A(g) \\ &= \dim_C S_G - \dim_C A_G. \end{aligned}$$

This gives the result.

3.63 REMARK. If V is real then $\lambda^n V$ is real. If V is quaternionic then $\lambda^n V$ is real for n even, quaternionic for n odd.

Proof. Suppose V admits a structure map j whose square is ϵ. Then V^n admits a structure map $j \otimes j \otimes \ldots \otimes j$ whose square is ϵ^n, and similarly for $\lambda^n V$.

3.64 REMARK. If V is real then $\psi^k V$ is real. If V is quaternionic then $\psi^k V$ is real for k even, quaternionic for k odd.

Proof. If we assign weight i to σ_i, then $P_k(\sigma_1, \sigma_2, \ldots, \sigma_k)$ is a polynomial of weight k. Now use 3.7.

We now move on to calculate $K_C(G \times H)$ in terms of $K_C(G)$ and $K_C(H)$. Let V be a G-space and W an H-space (over C). Then we can form $V \otimes W$, and make it a $G \times H$-space by

$$(g,h)(v \otimes w) = gv \otimes hw.$$

This defines a homomorphism of rings

$$\nu : K_C(G) \otimes K_C(H) \to K_C(G \times H).$$

3.65 THEOREM. The map ν is an isomorphism. More precisely, the inequivalent irreducible $G \times H$-spaces (over C) are precisely the products $V_i \otimes W_j$, where V_i runs over the inequivalent irreducible G-spaces and W_j over the inequivalent irreducible H-spaces.

Given the theorem for $K_C(G \times H)$, it is easy to locate the representations of $G \times H$ over R and Q; for an irreducible representation $V_i \otimes W_j$ is self-conjugate if and only if both V_i and W_j are self-conjugate; and then $V_i \otimes W_j$ is real or quaternionic according to the nature of V_i and W_j, as in 3.7.

Theorem 3.65 will follow immediately from the next two results.

3.66 LEMMA. If V is an irreducible G-space and W is an irreducible H-space (over C), then $V \otimes W$ is an irreducible $G \times H$-space.

3.67 LEMMA. Any $G \times H$-space U (over C) can be expressed in the form $\sum_{i,j} n_{ij} V_i \otimes W_j$. In particular, the irreducible $G \times H$-spaces have the form $V_i \otimes W_j$.

<u>First proof of 3.66</u>. We have

$$\chi_{V \otimes W}(g,h) = \chi_V(g) \cdot \chi_W(h).$$

Thus

$$\int_{(g,h)\in G\times H} \bar{\chi}_{V\otimes W}(g,h)\, \chi_{V\otimes W}(g,h)$$

$$= \int_{g\in G} \bar{\chi}_V(g)\chi_V(g) \int_{h\in H} \bar{\chi}_W(h)\chi_W(h)$$

$$= 1.$$

By 3.20 and 3.34, $V \otimes W$ is irreducible.

<u>Proof of 3.67</u>. By 3.25 we have an isomorphism over H:

$$\mu : \bigoplus_j \mathrm{Hom}_H(W_j, U) \otimes W_j \xrightarrow{\cong} U.$$

Let G act on $\mathrm{Hom}_H(W_j, U)$ by

$$(gk)w = g(kw) \quad \text{for} \quad k \in \mathrm{Hom}_H(W_j, U).$$

(It is easy to check that gk is an H-map.) Then μ is a $G \times H$-map. But by 3.20 we have an isomorphism of G-mocules

$$\mathrm{Hom}_H(W_j, U) \cong \bigoplus_i n_{ij} V_i.$$

Thus

$$U \cong \bigoplus_{i,j} n_{ij} V_i \otimes W_j.$$

Finally, if U is irreducible, it is clear that the sum can contain at most one factor.

Second proof of 3.66. Suppose that V and W are irreducible and $V \otimes W$ has a $G \times H$-subspace S. Then by 3.67 we have

$$S = \sum_{i,j} n_{ij} V_i \otimes W_j.$$

As H-spaces we have

$$V \otimes W \cong (\dim V) W;$$

so the only W_j which can have $n_{ij} \neq 0$ is W. Similarly, the only V_i which can have $n_{ij} \neq 0$ is V. Hence dim S is a multiple of $(\dim V)(\dim W)$, and $S = 0$ or $S = V \otimes W$.

We now move on to consider the case of a double covering $\pi : \tilde{G} \to G$. That is, π is an epimorphism of topological groups and $\mathrm{Ker}\,\pi = Z_2 = \{1, z\}$, say. Ker π is, of course, normal in $\tilde{G}$, and even central since Aut $Z_2 = 1$.

3.68 THEOREM. A character $\tilde{\chi} : \tilde{G} \to C$ factors as $\tilde{G} \xrightarrow{\pi} G \xrightarrow{\chi} C$ for a character χ if and only if $\tilde{\chi}$ factors as a map of sets. Moreover, χ is real if and only if $\tilde{\chi}$ is real; similarly, χ is quaternionic if and only if $\tilde{\chi}$ is quaternionic.

Proof. "Only if" is trivial. So suppose that $\tilde{V}$ is a representation of $\tilde{G}$. Then z acts on $\tilde{V}$ and satisfies $z^2 = 1$. So $\tilde{V}$ splits as the sum of the +1 and -1 eigenspaces of z, say $\tilde{V} = V \oplus V^-$. Since z is central, both V and V^- are $\tilde{G}$-spaces. We have

$$\tilde{\theta}(zg) = \theta(g) \oplus (-\theta^-(g))$$

and taking traces,

$$\tilde{\chi}(zg) = \chi(g) - \chi^-(g).$$

If $\tilde{\chi} : \tilde{G} \to C$ factors as a map of sets, then

$$\tilde{\chi}(zg) = \tilde{\chi}(g) = \chi(g) + \chi^-(g),$$

so $\chi^-(g) = 0$ and $V^- = 0$. Clearly $\tilde{V} = V$ is then a representation of G. If it carries a structure map commuting with the operations of $\tilde{G}$, then it carries the same structure map commuting with the operations of G.

3.69 REMARK. This theorem is also valid for virtual

characters.

3.70 EXERCISE. Extend 3.68 to any finite covering, assuming G compact connected and $\Lambda = C$.

We now turn to consider the representations of the torus.

3.71 PROPOSITION. If G is abelian and $\Lambda = C$ then every irreducible G-space V is one-dimensional.

Proof. For each $g \in G$ consider $\theta(g) : V \to V$. This is a G-map because G is abelian. By 3.22(ii), $\theta(g)$ is multiplication by some scalar $\lambda(g)$. So every subspace of V is stable under G and dim V = 1.

3.72 REMARK. In 3.71, $\lambda(g) \in C - \{0\}$.

3.73 REMARK. Suppose that G is a compact abelian group and V an irreducible G-space, so that θ may be written as $\lambda : G \to C - \{0\}$. Then $\lambda(G) \subset S^1 \subset C - \{0\}$, where S^1 is the unit circle in C.

First proof. If $|\lambda(g)| = r > 1$, then $|\lambda(g^n)| = r^n \to \infty$; and if $|\lambda(g)| = r < 1$, then $|\lambda(g^n)| = r^n \to 0$.

Second proof. Give V a positive definite Hermitian form H invariant under G. Then

$$H(v,v) = H(gv,gv) = |\lambda(g)|^2 H(v,v),$$

so

$$|\lambda(g)| = 1.$$

We recall that T^1 was defined to be R/Z.

3.74 PROPOSITION. A homomorphism $\alpha : T^1 \to T^1$ has the form $\alpha(x) = nx \bmod 1$ for some integer n.

Proof. By 2.11 and 2.13, or by the ordinary theory of covering spaces, α lifts to a homomorphism $\beta : R \to R$. Then $\beta(1) \equiv 0 \bmod 1$, so $\beta(1) = n \in Z$; and $\beta(a) = na$ for $a \in Z$, and $b\beta(a/b) = \beta(a) = na$ for $b \in Z$, so $\beta(a/b) = na/b$. By continuity, $\beta(x) = nx$ for all $x \in R$, and $\alpha(x) = nx \bmod 1$.

3.75 COROLLARY. A homomorphism $\alpha : T^k \to T^1$ has the form

$$\alpha(x_1, x_2, \ldots, x_k) = n_1 x_1 + \ldots + n_k x_k \bmod 1$$

for some $n_1, n_2, \ldots, n_k \in Z$.

3.76 COROLLARY. The irreducible complex T^k-spaces have the form

$$\lambda(x_1, x_2, \ldots, x_k) = \operatorname{Exp} 2\pi i(n_1 x_1 + \ldots + n_k x_k),$$

where $\operatorname{Exp} z = e^z$.

This follows from 3.71, 3.73 and 3.75, since T^1 is isomorphic to S^1 under $x \to \operatorname{Exp} 2\pi i x$.

For $1 \leq j \leq k$, let ρ_j be the T^k-space given by

$\lambda(x_1, x_2, \ldots, x_k) = \operatorname{Exp} 2\pi i x_j$.

Then ρ_j is invertible, and for any integers $n_1, n_2, \ldots, n_k$ (positive, negative or zero) $\rho_1^{n_1} \rho_2^{n_2} \ldots \rho_k^{n_k}$ is the T^k-space given by

$\lambda(x_1, x_2, \ldots, x_n) = \operatorname{Exp} 2\pi i(n_1 x_1 + \ldots + n_k x_k)$.

3.77 COROLLARY. $K_C(T^k)$ is the ring of finite Laurent series in $\rho_1, \rho_2, \ldots, \rho_k$, and so has no divisors of zero.

We have

$t(\rho_1^{n_1} \rho_2^{n_2} \ldots \rho_k^{n_k}) = \rho_1^{-n_1} \rho_2^{-n_2} \ldots \rho_k^{-n_k}$.

Thus the only irreducible representation of T^k which is self-conjugate is the trivial representation 1.

3.78 COROLLARY. The inequivalent irreducible real representation of T^k are

(i) the trivial representation 1 of dimension 1, and

(ii) the representations

$$r\left(\rho_1^{n_1}\rho_2^{n_2}\cdots\rho_k^{n_k}\right)=r\left(\rho_1^{-n_1}\rho_2^{-n_2}\cdots\rho_k^{-n_k}\right)$$

for $(n_1, n_2, \ldots, n_k) \neq (0, 0, \ldots, 0)$, which are of dimension 2.

This follows from the above by the discussion of 3.57.

Chapter 4

MAXIMAL TORI IN LIE GROUPS

Notice. From 4.5 onwards, G will be a compact connected Lie group.

4.1 DEFINITION. Let G be a topological group and let $g \in G$. Let H be the subgroup generated by g. Then g is a generator of G if cl H = G, where cl denotes the closure.

G is monogenic (or monothetic) if it has a generator.

4.2 EXERCISE. Monogenic implies Abelian.

4.3 PROPOSITION. The torus T^k is monogenic. Indeed, generators are dense in T^k.

Proof. Let $U_1, U_2, \ldots$ be a countable base for the open sets of T^k. Let $T^k = R^k/Z^k$ have co-ordinates $(x_1, \ldots, x_k)$.

Then a cube is a set $\{x \in T^k;\ |x_i - \xi_i| \leq \epsilon\}$ for some fixed point ξ, and real $\epsilon > 0$. Let C_0 be any cube. Then we will define a descending sequence of subcubes whose intersection will be a generator.

Suppose, inductively, that we have defined $C_0 \supset C_1 \supset \ldots \supset C_{m-1}$ and that C_{m-1} has side 2ϵ. Then there is an integer $N(m)$ such that $N \cdot 2\epsilon > 1$, so that the image of C_{m-1} under multiplication by N is T^k. We can find $C_m \subset C_{m-1}$ such that $N \cdot C_m \subset U_m$.

Let $g \in \cap_m C_m$. Then $g^{N(m)} \subset U_m$, so g is a generator of T^k.

4.4 PROPOSITION. Let G be an Abelian topological group, with $T^k \subset G$ such that $G/T^k = Z_m$. Then G is monogenic.

Proof. Let t be a generator of T^k. Choose $u \in G$ to project to a generator of Z_m. Then $mu \in T^k$ and $t - mu \in T^k$. T^k is divisible, so there is $s \in T^k$ with $ms = t - mu$. Take $g = u + s$. Then $mg = m(u + s) = t$, so the powers of g are dense in T. Translating by rg, the powers of g are dense in the coset of T containing ru. This gives all cosets.

4.5 NOTICE. From now on G is a compact connected Lie group.

4.6 DEFINITION. A <u>maximal torus</u> $T \subset G$ is:

(i) a subgroup which is a torus, such that

(ii) if $T \subset U \subset G$ and U is a torus then $T = U$.

4.7 REMARK. If G is not compact, it need not have any non-trivial tori.

4.8 PROPOSITION. Any subtorus of G is contained in a maximal torus.

<u>Proof</u>. Consider a strictly increasing sequence of subtori $T_1 \subset T_2 \subset \ldots \subset G$. Then $L(T_1) \subset L(T_2) \subset \ldots \subset L(G)$ is a strictly increasing sequence, and so is finite.

4.9 PROPOSITION. Let T be a maximal torus of G, and A a connected Abelian subgroup of G with $T \subset A$. Then $T = A$.

<u>Proof</u>. $T \subset A \subset \mathrm{cl}\, A$. But cl A is a closed connected Abelian subgroup, and is therefore a torus (2.20). Thus $T = \mathrm{cl}\, A$ and $T = A$.

4.10 CONSTRUCTIONS. If T is a torus of G, it operates on G_e by $T \subset G \xrightarrow{Ad} \text{Aut } G_e$. Choose a positive definite symmetric form on G_e invariant under G, and so under T. Then (3.78) G_e splits into orthogonal irreducible T-spaces of dimensions 1 and 2. Those of dimension 1 are trivial. We can choose an orthonormal base in those of dimension 2, and represent T by $T \to SO(2)$.

4.11 DEFINITION. The <u>integer lattice</u> of L(T) is $\exp^{-1}(e)$ where $\exp : L(T) \to T$.

4.12 PROPOSITION. $L(G) = G_e$ splits as a T-space in the form $V_0 \oplus \Sigma_1^m V_i$, where T acts on V_0 trivially, dim $V_i = 2$ for $i > 0$ and T acts on V_i as

$$\begin{bmatrix} \cos 2\pi\theta_i(t) & -\sin 2\pi\theta_i(t) \\ \sin 2\pi\theta_i(t) & \cos 2\pi\theta_i(t) \end{bmatrix}.$$

Here $\theta_i : T \to R/Z$ is given by a linear form $\theta_i : L(T) \to R$ taking integer values on the integer lattice, and no θ_i is zero.

4.13 DEFINITION. If T is a maximal torus, the functions $\pm\theta_i$ are called the <u>roots</u> of G. By 3.24 they are well defined in terms of T. We will see that they are independent of T.

4.14 PROPOSITION. T is maximal if and only if $V_0 = L(T)$.

Proof. It is clear that $L(T) \subset V_0$.

(i) Suppose $V_0 = L(T)$ and $T \subset T'$. Then $L(T) \subset L(T') \subset V_0' \subset V_0$, so $L(T) = L(T')$ and $T = T'$.

(ii) Suppose $V_0 \neq L(T)$. Then there is $X \in V_0$, $X \notin L(T)$. Now $\exp(tX)$, for $t \in R$, is a 1-parameter subgroup H of G on which T acts trivially, and which is not contained in T. Therefore the subgroup generated by T and H is a connected Abelian subgroup strictly containing T, so T is not maximal.

4.15 COROLLARY. dim G - dim T is even.

4.16 EXAMPLE. Let $G = U(n)$, and let T be the set of diagonal matrices:

$$D = \begin{bmatrix} \exp 2\pi i x_1 & & \\ & \ddots & \\ & & \exp 2\pi i x_n \end{bmatrix}.$$

$LU(n)$ can be decomposed into the following summands.

(i) Matrices $\begin{bmatrix} id_1 & & \\ & \ddots & \\ & & id_n \end{bmatrix}$ with d_j real.

This is L(T).

(ii) Matrices

$$M_{rs} = \begin{array}{c} \\ r \\ \\ s \\ \end{array}\begin{bmatrix} & r & & s & \\ & & & \xi & \\ & & & & \\ & -\bar{\xi} & & & \\ & & & & \end{bmatrix}$$

for r < s. Then

$$DM_{rs}D^{-1} = \begin{bmatrix} & & & & \\ & & & w & \\ & & & & \\ & -\bar{w} & & & \\ & & & & \end{bmatrix}$$

where $w = \exp[2\pi i(x_r - x_s)]z$ and $\theta_{rs} = x_r - x_s$.

The matrices (i) and (ii) generate L(G), so $V_0 = L(T)$ and T is maximal. The roots are $(x_r - x_s)$.

4.17 EXAMPLE. Let G = SU(n). Then the matrices M_{rs} of the previous example are in LSU(n), since the derivative with respect to t of $|I + tM_{rs}|$ at t = 0 is zero. Similarly, matrices of type (i) with $\Sigma d_i = 0$ are in LSU(n).

Let T be the set of diagonal matrices

$$D = \begin{bmatrix} \exp 2\pi i x_1 & & \\ & \ddots & \\ & & \exp 2\pi i x_n \end{bmatrix}$$

with $\Sigma x_i = 0$. The functions $(x_r - x_s)$ are still nontrivial, so $V_0 = L(T)$, T is maximal, and the roots are $(x_r - x_s)$.

4.18 EXAMPLE. Let $G = Sp(n)$, and let T be the set of diagonal matrices

$$D = \begin{bmatrix} \exp 2\pi i x_1 & & \\ & \ddots & \\ & & \exp 2\pi i x_n \end{bmatrix}.$$

L Sp(n) splits into the following summands.

(i) Matrices $\begin{bmatrix} id_1 & & \\ & \ddots & \\ & & id_n \end{bmatrix}$ with d_i real.

(ii) Matrices

$$M_{rs} = \begin{array}{c} \\ r \\ \\ s \end{array}\begin{bmatrix} & r & & s & \\ & & & z & \\ & & & & \\ & -\bar{z} & & & \\ & & & & \end{bmatrix}$$

with $z \in C$.

(iii) Matrices

$$N_r = \begin{array}{c} \\ r \\ \\ \end{array}\begin{bmatrix} & \overset{r}{} & \\ & zj & \\ & & \end{bmatrix}$$

with $z \in C$. Here

$$DN_rD = \begin{bmatrix} & & \\ & \exp(2\pi i x_r) zj \exp(-2\pi i x_r) & \\ & & \end{bmatrix}$$

$$= \begin{bmatrix} & & \\ & \exp(4\pi i x_r) zj & \\ & & \end{bmatrix}$$

(iv) Matrices

$$P_{rs} = \begin{array}{c} \\ r \\ \\ s \\ \end{array}\begin{bmatrix} \overset{r}{} & & \overset{s}{} \\ & & zj \\ & & \\ zj & & \end{bmatrix}$$

with $z \in C$. Here

$$DP_{rs}D^{-1} = \begin{bmatrix} & \exp 2\pi i(x_r + x_s)zj \\ \exp 2\pi i(x_r + x_s)zj & \end{bmatrix}.$$

Thus $V_0 = L(T)$, T is a maximal torus, and the roots are

$\pm 2x_r$, $(x_r - x_s)$ and $\pm(x_r + x_s)$ for $r \neq s$.

4.19 EXAMPLE. Let $G = SO(2n)$. We have $U(n) \subset SO(n)$.

Take T to be the image of the maximal torus we had in U(n).

That is, T is the set of matrices

$$D = \begin{bmatrix} D_1 & & & \\ & D_2 & & \\ & & \ddots & \\ & & & D_n \end{bmatrix}$$

where $D_i = \begin{bmatrix} \cos 2\pi x_i & -\sin 2\pi x_i \\ \sin 2\pi x_i & \cos 2\pi x_i \end{bmatrix}.$

Now LSO(2n) splits into the following summands.

(i) L(T), consisting of matrices

$$\begin{bmatrix} \begin{matrix} 0 & -d_1 \\ d_1 & 0 \end{matrix} & & \\ & \ddots & \\ & & \begin{matrix} 0 & -d_n \\ d_n & 0 \end{matrix} \end{bmatrix}.$$

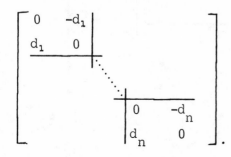

(ii) The rest of LU(n), consisting of matrices

$$M_{rs} = \begin{array}{c} \\ r \\ \\ s \\ \\ \end{array} \overset{\begin{array}{cc} r & \qquad s \end{array}}{\left[\begin{array}{cc} & W \\ & \\ -W^T & \\ & \end{array}\right]}$$

where $W = \begin{bmatrix} x & -y \\ y & x \end{bmatrix}$.

Then $DM_{rs}D^{-1} = M'_{rs}$ with

$$\begin{bmatrix} x' \\ y' \end{bmatrix} = \begin{bmatrix} \cos 2\pi(x_r - x_s) & -\sin 2\pi(x_r - x_s) \\ \sin 2\pi(x_r - x_s) & \cos 2\pi(x_r - x_s) \end{bmatrix} \begin{bmatrix} x \\ y \end{bmatrix}.$$

So T acts with $\theta_{rs} = x_r - x_s$.

(iii) Let

$$E_s = \begin{array}{c} \\ \\ \\ s \\ \\ \\ \\ \end{array} \overset{s}{\left[\begin{array}{ccccccc} 1 & & & & & & \\ & \ddots & & & & & \\ & & 1 & & & & \\ & & & \begin{array}{|cc|}\hline 1 & 0 \\ 0 & -1 \\ \hline \end{array} & & & \\ & & & & 1 & & \\ & & & & & \ddots & \\ & & & & & & 1 \end{array}\right]}$$

and take matrices $E_s M_{rs} E_s^{-1}$. In this case, T acts with

$\theta_{rs} = x_r + x_s$.

Thus $V_0 = L(T)$, T is a maximal torus, and the roots are $(x_r - x_s)$, $\pm (x_r + x_s)$ for $r \neq s$.

4.20 EXAMPLE. Let $G = SO(2n + 1)$. We have $SO(2n) \subset SO(2n + 1)$ by letting $SO(2n)$ act on the first 2n co-ordinates. Let T be the maximal torus we had in $SO(2n)$.

Then $LSO(2n + 1)$ splits into the following summands.

(i) $LSO(2n)$.

(ii) Matrices

$F_r =$

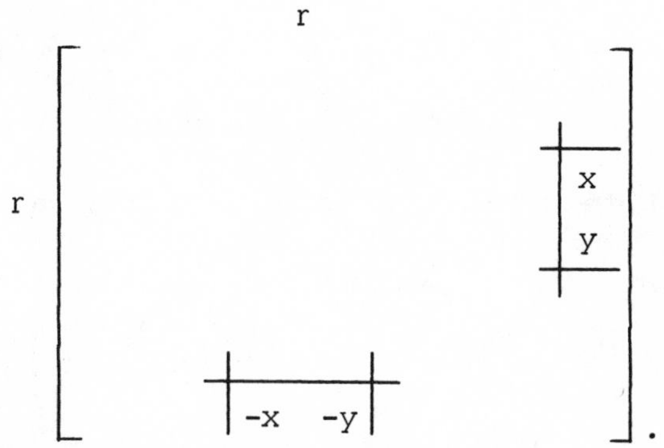

Here D acts by rotation through x_r.

Thus $V_0 = L(T)$, T is a maximal torus, and the roots are $\pm x_r$, $(x_r - x_s)$ and $\pm(x_r + x_s)$ for $r \neq s$.

4.21 THEOREM. Let $T \subset G$ be a maximal torus. Then any $g \in G$ is contained in a conjugate of T.

Proof. (Following A. Weil [21]; see also [11].)

Consider the left coset space G/T, and let $f : G/T \to G/T$ be induced by left multiplication by g, that is, $f(xT) = gxT$. Then a fixed point of f is a coset xT with $gxT = xT$, that is, $g \in xTx^{-1}$. So we only need to show that f has a fixed point. We will use the form of Lefschetz's fixed point theorem given by A. Dold [6]. (This theorem applies to manifolds rather than simplicial complexes). We summarise what we need:

Let $f : X \to X$ be a continuous map, and define $\Lambda(f) \in Z$ by taking $f^* : H^q(X; Q) \to H^q(X; Q)$ and setting $\Lambda(f) = \sum_q (-1)^q \operatorname{Tr} f^*$. Then $\Lambda(f)$ depends only on the homotopy class of f. If f has no fixed points, then $\Lambda(f) = 0$. If f has only isolated fixed points (and so a finite number of fixed points), then $\Lambda(f)$ is the number of fixed points counted with multiplicity, which is defined as follows. Let X be a smooth manifold and x a fixed point of f. Consider $1 - f' : X_x \to X_x$. If $\det(1 - f') > 0$, then f has multiplicity $+1$ at x: if $\det(1 - f') < 0$, then the multiplicity is -1. We do not need to discuss the case $\det(1 - f') = 0$.

To compute $\Lambda(f)$ we may replace f with any homotopic map f_0. So we may replace g with any other $g_0 \in G$, since G is path-connected. Take g_0 to be a generator of T (4.1), and let f_0 be the corresponding map. Then the fixed points of f_0 are the cosets nT for n in N(T), the normaliser of T in G (as the reader will easily verify). Let us examine N(T).

N(T) is a closed subgroup of G, and so is a Lie group (2.27, 2.26), and the identity component $N(T)_1$ is open and so has only a finite number of cosets. Now $N(T)_1 = T$, which we see as follows. N(T) acts on T by conjugation (i.e., $n(t) = ntn^{-1}$) and Aut T is discrete, so $N(T)_1$ acts trivially. (The reader should verify that $N \to \text{Aut } T$ is continuous with this topology on Aut T. Note that this map arises from the map $N \times T \to T$ which is a restriction of the map $G \times G \to G$ given by $(g,h) \to ghg^{-1}$). If $N(T)_1$ properly contains T it contains a 1-parameter subgroup not contained in T but computing with T, contradicting the maximality of T. It follows that $N(T)_1 = T$, that T has only a finite number of cosets in N(T), and that f_0 has only a finite number of fixed points.

It suffices to consider just one of these fixed points, say T, as follows. Let nT be another fixed point. Define $r_n : G/T \to G/T$ by $r_n(gT) = gTn$. This is a well-defined diffeomorphism, commutes with f_0, and takes T to nT. Thus the multiplicity at nT is the same as at T.

Observe that f_o can also be defined as $f_o(xT) = g_o x g_o^{-1} T$. That is, f_o is obtained as a quotient of the map $G \to G$ given by $x \to g_o x g_o^{-1}$. This has the merit that e goes to e.

To obtain a basis of $(G/T)_T$, take a basis for T_e, extend it to a basis of G_e, and discard the vectors of T_e. Then (4.12, 4.14) $1 - f_o'$ has the form

$$\left[\begin{array}{cc|c} 1 - \cos 2\pi\theta_1(g_o) & \sin 2\pi\theta_1(g_o) & \\ -\sin 2\pi\theta_1(g_o) & 1 - \cos 2\pi\theta_1(g_o) & 0 \\ \hline & & \ddots \\ \multicolumn{2}{c|}{0} & \end{array}\right]$$

Therefore $\det(1 - f_o') = \Pi_1^m \begin{vmatrix} 1 - \cos 2\pi\theta_1(g_o) & \sin 2\pi\theta_1(g_o) \\ -\sin 2\pi\theta_1(g_o) & 1 - \cos 2\pi\theta_1(g_o) \end{vmatrix}$

which is greater than 0 unless $\cos 2\pi\theta_r(g_o) = 1$ for some r. But $\theta_r(g_o) \not\equiv 0 \bmod 1$, since θ_r is a nontrivial function on T (4.12). Hence the multiplicity is +1, and $\Lambda(f) = |N(T)/T| > 0$. Thus f has at least one fixed point, and the theorem is proved.

4.22 COROLLARY. Every element of G lies in a maximal torus, since the conjugate of a maximal torus is a maximal torus.

4.23 COROLLARY. Any two maximal tori, T, U are conjugate.

Proof. Let u be a generator of U. Then $u \in xTx^{-1}$ for some $x \in G$, and thus $U \subset xTx^{-1}$. But U is a maximal torus, so $U = xTx^{-1}$.

Hence any construction apparently dependent on a choice of T is independent of the choice up to an inner automorphism of G.

4.24 DEFINITION. It follows that any two maximal tori have the same dimension. This is called the rank of G, and written k or l.

4.25 PROPOSITION. Let S be a connected Abelian subgroup of G, and let $g \in G$ commute with all elements of S. Then there is a torus T containing g and S.

Proof. Let H be the subgroup generated by g and S. H is Abelian, so Cl H is a compact Abelian Lie group. Therefore the identity component $(\mathrm{Cl}\ H)_1$ is a torus. $\mathrm{Cl}\ H/(\mathrm{Cl}\ H)_1$ is finite and generated by g, so $\mathrm{Cl}\ H/(\mathrm{Cl}\ H)_1 \cong Z_m$ for some integer m. By 4.4, Cl H has a generator h which lies in some maximal torus T. Then $g \cup S \subset H \subset \mathrm{Cl}\ H \subset T$.

4.26 PROPOSITION. Let T be a maximal torus of G. If $T \subset A \subset G$ where A is Abelian, then $T = A$. That is, a maximal torus is a maximal Abelian subgroup.

Proof. Let $g \in A$. Then (4.25) there is a torus U containing g and T. But T is maximal so $U = T$, and $g \in T$. Thus $A \subset T$.

4.27 EXAMPLE. If $a \in U(n)$ commutes with all diagonal matrices it is itself diagonal.

4.28 REMARK. It is not, in general, true that a maximal Abelian subgroup is a torus. For example, let $G = SO(n)$ and consider the set of matrices of the form

$$\begin{bmatrix} \pm 1 & & \\ & \ddots & \\ & & \pm 1 \end{bmatrix}.$$

These form a maximal Abelian subgroup.

4.29 DEFINITION. Let T be a maximal torus of G. Then the Weyl group W (or Φ) of G is the group of automorphisms of T which are the restrictions of inner automorphisms of G. This is independent of the choice of T.

Any such automorphism has the form $t \to ntn^{-1}$, $n \in N(T)$.

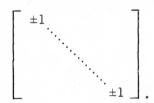

N(T) is a closed subgroup of G, and so compact. Let Z(T) be the centraliser of T, that is, the set of $z \in G$ such that $ztz^{-1} = t$ all $t \in T$. Z(T) is also closed, and $T \subset Z(T) \subset N(T)$. Thus N(T) maps onto $N(T)/Z(T) \cong W$. N(T)/T is finite (see the proof of 4.21), so W is finite.

Since we are considering G connected, Z(T) = T (4.25), and W = N(T)/T.

4.30 COROLLARY of 4.21. Let V be a G-space. Then χ_V is determined by its restriction to T and is invariant under W.

4.31 COROLLARY. The homomorphism $i^* : K(G) \to K(T)$ of (complex) representation rings is mono, and its image is contained in the subring of elements invariant under W.

4.32 PROPOSITION. Restriction gives a one-one correspondence between class functions on G and continuous functions on T invariant under W.

Proof. We have already shown that the correspondence is mono.

Suppose given $f : T \to Y$ continuous and invariant under W. Extend f to $\bar{f} : G \to Y$ by $\bar{f}(xtx^{-1}) = f(t)$. To show that $\bar{f}$ is well-defined we need:

4.33 LEMMA. If $t_1, t_2 \in T$ are conjugate in G, then there is $w \in W$ with $t_2 = wt_1$.

Proof. Let $H = N(t_2) = Z(t_2)$ and let $t_2 = gt_1g^{-1}$. Then $T \subset Z(t_2)$ and, since $T \subset Z(t_1)$, $gTg^{-1} \subset Z(t_2)$ also. H is a closed subgroup of G, and so a Lie group, and so T, gTg^{-1} are maximal tori of H. Therefore there is $h \in H_1$ such that $T = hgTg^{-1}h^{-1}$, where H_1 is the identity component of H. But $h \in Z(t_2)$ so $hgt_1gh^{-1} = t_2$. Thus conjugation by hg, which is in W, sends t_1 to t_2.

Completion of 4.32. It remains to check that $\bar{f}$ is continuous.

Well, suppose that $\bar{f}$ is not continuous. Then there is a sequence $g_n \to g_\infty$ such that no subsequence of $\bar{f}g_n$ tends to $\bar{f}g_\infty$. Let $g_n = x_n t_n x_n^{-1}$ and take a subsequence with $x_n \to x_\infty$, $t_n \to t_\infty$ for some x_∞, t_∞. Then $g_n \to x_\infty t_\infty x_\infty^{-1}$, and so $x_\infty t_\infty x_\infty^{-1} = g_\infty$. Then $\bar{f}(g_n) = f(t_n) \to f(t_\infty) = \bar{f}(g_\infty)$, which contradicts our hypothesis. Thus 4.32 is proved.

4.34 LEMMA. Let $N(g)_1$ be the identity component of the normaliser of some $g \in G$. Then $N(g)_1$ is the union of the maximal tori of G containing g.

Proof. Clearly $N(g)_1$ contains all such tori. So let $n \in N(g)_1$. Then n lies in a maximal torus S of $N(g)_1$. S commutes with g, so (4.25) there is a maximal torus T of G containing S and g.

4.35 COROLLARY. The following two definitions are equivalent:

(i) $g \in G$ is regular if it is contained in just one maximal torus,

singular if it is contained in more than one maximal torus,

(ii) $g \in G$ is regular if $\dim N(g) = \text{rank } G$,

singular if $\dim N(g) > \text{rank } G$.

Proof. If g lies in just one T, then $\dim N(g) = \dim N(g)_1 = \dim T$. If g lies in T_1 and T_2, and $T_1 \neq T_2$, then $LT_1 \neq LT_2$ and $LN(g) \supset LT_1 + LT_2$ so $\dim N(g) > \dim T$.

4.36 EXAMPLE. Let $G = Sp(1)$, which is the set of quaternions q with $|q| = 1$. Maximal tori are circles $\cos\theta + p\sin\theta$, for p any pure imaginary quaternion with $|p| = 1$.

The singular points are ± 1, with $\dim N(\pm 1) = 3$.

All other points g are regular, and $\dim N(g) = 1$.

4.37 PROPOSITION. W permutes the roots of G.

<u>Proof</u>. (The notation was introduced in 1.10.) For each $w \in W$ we must consider two representations for T, namely $T \xrightarrow{Ad} \mathrm{Aut}\, G_e$ and $T \xrightarrow{w} T \xrightarrow{Ad} \mathrm{Aut}\, G_e$. It will suffice to show that these are equivalent. But $w = A_x | T$ for some $x \in G$, and then $G_e \xrightarrow{A_x'} G_e$ is the required equivalence, since

$$\begin{array}{ccc} T & \xrightarrow{A_x} & T \\ {\scriptstyle Ad}\downarrow & & \downarrow{\scriptstyle Ad} \\ \mathrm{Aut}\, G_e & \longrightarrow & \mathrm{Aut}\, G_e \end{array}$$

is commutative, where the bottom map is induced from A_x'.

4.38 DEFINITION. Let $U_r = \{t \in T ; \theta_r(t) \equiv 0 \bmod 1\}$. U_r is a closed subgroup of T of dimension $k - 1$, where $k = \operatorname{rank} G$. It is clearly monogenic. It need not be connected. For instance:

4.39 EXAMPLE. In Sp(1), $\theta_1 = 2x_1$ and U_1 is given by $x_1 \equiv 0$ or $\frac{1}{2}$ mod 1.

4.40 LEMMA. If t lies in exactly ν of the U_r, then $\dim N(t) = k + 2\nu$.

Proof. Let $V \subset L(G)$ be the subspace on which t acts as the identity. Then, by definition, $\dim V = k + 2\nu$. We show that $N(t)_e = V$.

(i) The elements of N(t) commute with t, so t acts as the identity on N(t) and so on $N(t)_e$. Thus $N(t)_e \subset V$.

(ii) Suppose $x \in V$. Then t acts trivially on x, and so on the 1-parameter subgroup H corresponding to x. Therefore $H \subset N(t)$ and $x \in N(t)_e$. Thus $V \subset N(t)_e$.

4.41 COROLLARY. $t \in T$ is regular if it lies in no U_r, and singular if it lies in some U_r.

4.42 COROLLARY. The singular elements of G form a set of dimension $\leq n - 3$, where $n = \dim G$, in the sense that this set is the image of a compact manifold of dimension $n - 3$ under a smooth map.

Proof. Let u be a generator of U_r. Then $\dim N(u) \geq k + 2$, and, if $z \in N(u)$, z fixes each power of u and so fixes every element of U_r.

Define a map $f : G/N(u) \times U_r \to G$ by $f(g,t) = gtg^{-1}$. Then Imf consists of all points in conjugates of U_r, f is smooth, and $\dim G/N(u) \times U_r \leq n - (k + 2) + (k - 1) = n - 3$. All the singular points are obtained with r running over a finite set. Hence the result.

Chapter 5

GEOMETRY OF THE STIEFEL DIAGRAM

(Note: This is not the Dynkin-Coxeter diagram.)

Notice. Throughout this chapter G is a compact connected Lie group, and T is a maximal torus of G.

5.1 DEFINITION. The infinitesimal diagram of G is the figure in L(T) consisting of the hyperplanes $L(U_r)$.

The diagram of G is the figure in L(T) consisting of the hyperplanes given by $\theta_r(t) \in Z$. This is the inverse image under exp of the singular points of G in T.

5.2 EXAMPLES of diagrams.

(i) U(2). Root $x_1 - x_2$.

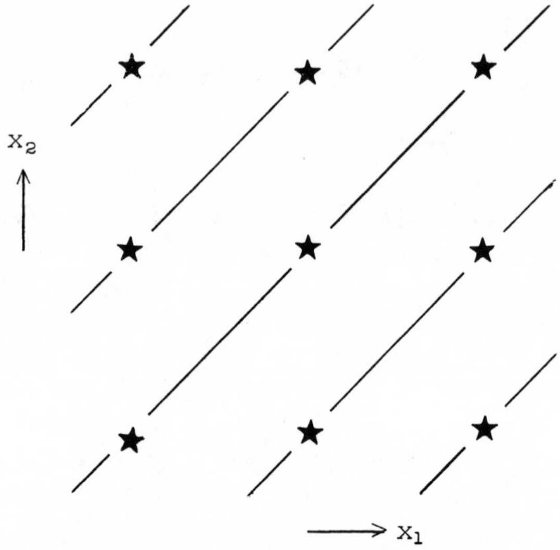

The integer lattice is marked with asterisks.

(ii) SO(4). Roots $x_1 \pm x_2$.

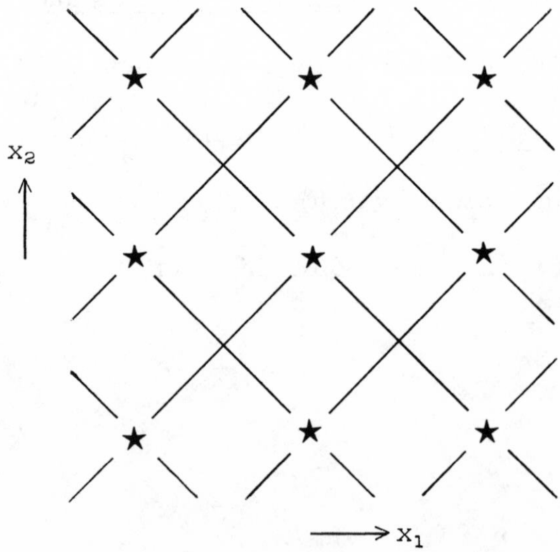

(iii) SO(5). Roots $x_1 \pm x_2$, x_1, x_2.

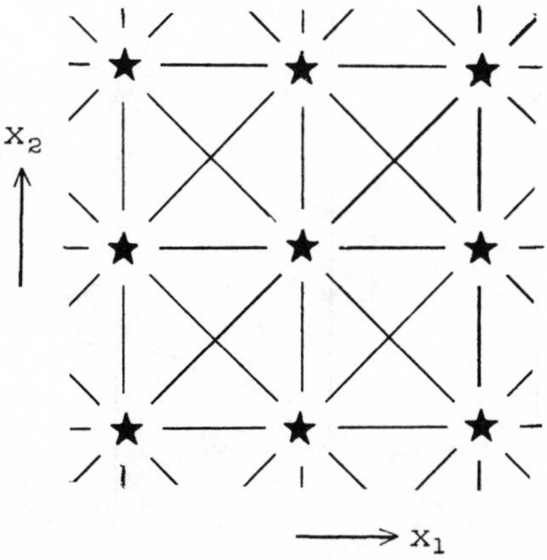

(iv) Sp(2). Roots $x_1 \pm x_2$, $2x_1$, $2x_2$.

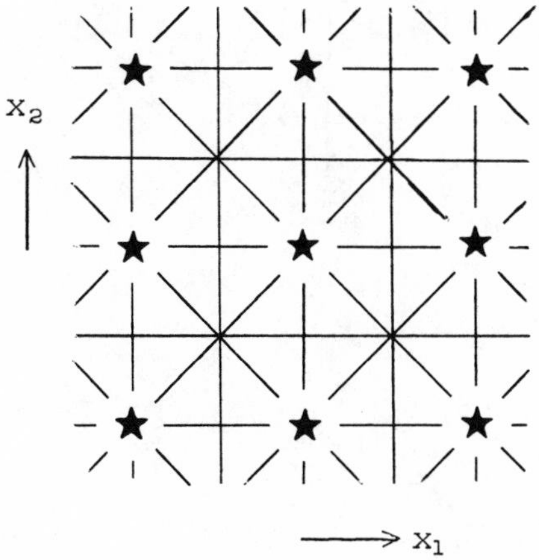

(v) SU(3). Roots $x_1 - x_2$, $x_2 - x_3$, $x_3 - x_1$.

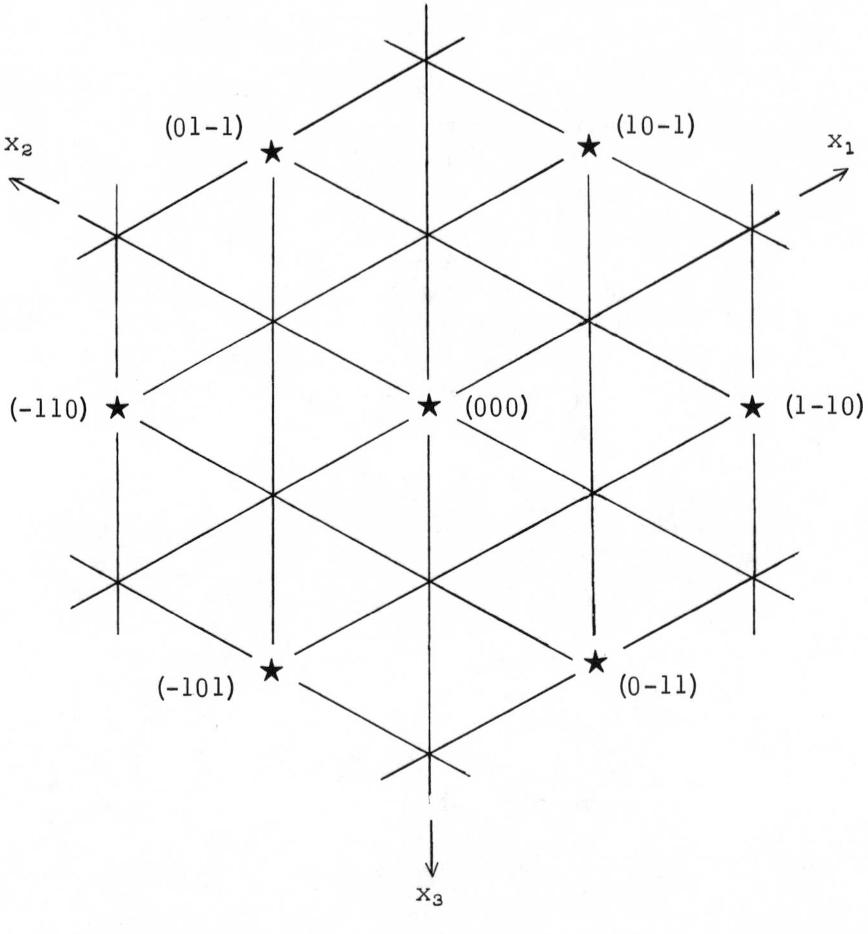

5.3 PROPOSITION. $Z(G) = \cap U_r$.

Proof. Firstly, $Z(G) \subset Z(T) = T$.

Now, if $z \in Z(G)$, z acts trivially on G and so on G_e. Therefore $\theta_r(z) \equiv 0 \bmod 1$ for each r and $z \in \cap_r U_r$.

Conversely, if $g \in T$ and $\theta_r(g) \equiv 0 \bmod 1$ for each r then g acts trivially on G_e, and so trivially on G (2.17).

5.4 EXAMPLES.

(i) U(n). $\cap U_r$ is given by $x_1 \equiv \ldots \equiv x_n \bmod 1$, so the centre consists of matrices $e^{2\pi i x} I$.

(ii) SU(n). $\cap U_r$ is given by $x_1 \equiv \ldots \equiv x_n \bmod 1$ and $x_1 + \ldots + x_n \equiv 0 \bmod 1$. Thus the centre consists of matrices ωI where $\omega^n = 1$.

(iii) Sp(n). $\cap U_r$ is given by $x_i \pm x_j \equiv 0 \bmod 1$ all i,j, i.e., $x_i \equiv 0 \bmod 1$ all i or $x_i \equiv \frac{1}{2} \bmod 1$ all i. Thus the centre consists of matrices $\pm I$.

(iv) SO(2n). $\cap U_r$ is given by $x_i \pm x_j \equiv 0 \bmod 1$ for $i \neq j$. For $n > 1$, this is the same as for Sp(n), and the centre consists of $\pm I$. Of course, SO(2) is Abelian.

(v) SO(2n + 1). $\cap U_r$ is given by $x_r \equiv 0 \bmod 1$ all r. Thus the centre consists of just the identity matrix I.

5.5 THEOREM. If $r \neq s$ then θ_r and θ_s are linearly independent.

Proof. U_r has dimension k - 1. We show that

dim $N((U_r)_1) = k + 2$. The result will then follow from 4.40 applied to a generator of $(U_r)_1$. We need two lemmas.

5.6 LEMMA. Suppose $H \subset T$, and that H is a closed subgroup which is normal in G. Then

(i) $N(T/H) = N(T)/H$.

(ii) T/H is a maximal torus in G/H.

(iii) $W(G/H) \cong W(G)$.

Proof.

(i) If n preserves T then nH preserves T/H. Conversely, if $n(tH)n^{-1} \subset T$ then $ntn^{-1} \subset T$.

(ii) T/H is a compact connected Abelian subgroup of G/H, and so a torus.

Now suppose $T/H \subset U/H$, where U/H is a torus in G/H. Then $U/H \subset N(T/H) = N(T)/H$, so $T \subset U \subset N(T)$. Therefore $\dim T = \dim U$, so $\dim T/H = \dim U/H$ and $T/H = U/H$.

(iii)
$$W(G/H) = N(T/H) \big/ T/H = N(T)/H \big/ T/H$$
$$\cong N(T)/T$$
$$\cong W(G).$$

5.7 LEMMA. If $\dim T = 1$ then

(i) $n = 1$ and $W = 0$, or

(ii) $n = 3$ and $W = Z_2$.

[Note: In fact in (i) $G = S^1$, and in (ii) $G = SO(3)$ or $Sp(1)$.]

Proof. If $n = 1$ then clearly $G = T = S^1$ and $W = 0$. So suppose $n > 1$.

Take an invariant norm in $L(G)$ and let v be a unit vector in $L(T)$. Define $f : G/T \to S^{n-1} \subset L(G)$ by $f(g) = (\mathrm{Ad}\, g)v$. Then f is well-defined, continuous (even smooth) and is mono for, if $(\mathrm{Ad}\, g_1)v = (\mathrm{Ad}\, g_2)v$ then $\mathrm{Ad}(g_1^{-1}g_2)v = v$, so $g_1^{-1}g_2$ fixes v and therefore fixes T. It follows that $g_1^{-1}g_2 \in T$ and $g_1T = g_2T$.

Now G/T is compact and S^{n-1} Hausdorff, so f is a homeomorphism of G/T with its image in S^{n-1}. But G/T and S^{n-1} are both compact manifolds of dimension $(n - 1)$, so f is onto. Then there exists $g \in G$ such that $(\mathrm{Ad}\, g)v = -v$, and therefore g acts on T by $gtg^{-1} = t^{-1}$. Now T has only two automorphisms, so $W = Z_2$.

Let i be the generator of $\pi_1(T)$. Since G is connected, g can be joined to e by an arc in G. So, in $\pi_1(G)$, $i = -i$, that is, $2i = 0$.

Now we have in fact (2.37) a fibration $S^1 \to G \to G/T \cong S^{n-1}$. From the exact homotopy sequence we have that $\pi_2(S^{n-1}) \to \pi_1(S^1) \to \pi_1(G)$ is exact. But

$\pi_1(S^1) \to \pi_1(G)$ is not mono, since $2i \to 0$. So $\pi_2(S^{n-1}) \neq 0$ and $n = 3$.

Proof of 5.5. Consider $(U_r)_1$, the identity component of U_r. This is a torus of dimension $k - 1$. Let u be a generator. We wish to show that $u \notin U_s$ for $r \neq s$, for then θ_s will not be a multiple of θ_r.

Consider $N(u)_1$. T is a maximal torus of $N(u)_1$. The elements of $N(u)$ fix u and so fix every element of $(U_r)_1$. We can apply 5.6 with $N(U)_1$ as G, T as T, and $(U_r)_1$ as H. Then $T/(U_r)_1$ is a maximal torus in $N(u)_1/(U_r)_1$, and $W(N(u)_1/(U_r)_1) \cong W(N(u)_1)$. Now $T/(U_r)_1$ has dimension 1, so (5.7) $N(u)_1/(U_r)_1$ has dimension 1 or 3, and $N(u)_1$ has dimension k or $k + 2$. But (4.40) $N(u)_1$ has dimension $k + 2\nu$ where u lies in exactly ν of the U_r. Hence $\nu = 1$ and u does not lie in U_s.

5.8 THEOREM. For each r there is an element $\varphi_r \in W$ which is not the identity but which leaves every point of U_r fixed.

Proof. We use the same proof as 5.5, but with a different choice of u.

Consider U_r. We observed (4.38) that U_r is monogenic. Let v be a generator.

Now consider $N(v)_1$. T is a maximal torus of $N(v)_1$, and $N(v)_1$ fixes every element of U_r. We can apply 5.6 with $N(v)_1$ as G, T as T, and U_r as H. We deduce that T/U_r is a maximal torus in $N(v)_1/U_r$, and $N(v)_1/U_r$ has dimension 1 or 3. By 4.40, $\dim N(v_1)/U_r \geq 3$, so $\dim N(v_1)/U_r = 3$ and $W(N(v)_1/U_r) \cong Z_2$. That is, there is $n \in N(v)_1$ which fixes each point of U_r and which maps T/U_r by $t \to t^{-1}$.

5.9 COROLLARY (of the proof). φ_r is the inner automorphism induced by an element n which can be joined to e by a path of which each point leaves each point of U_r fixed.

5.10 COROLLARY. U_r has either one or two components.

Proof. φ_r acts on $T/(U_r)_1$ by $t \to t^{-1}$, which has only two fixed points, namely 0 and $\frac{1}{2}$ mod 1. But $U_r/(U_r)_1$ is fixed by φ_r.

5.11 EXAMPLE. The root $2x_r$ of Sp(n) gives U_r with two components.

5.12 DEFINITION. For each r let $\epsilon_r = \pm 1$. Consider the set

$$\{t \in L(T);\quad \epsilon_r\theta_r(t) > 0 \quad \text{all } r\}.$$

This is either empty or is a non-empty convex set. In the latter case it is called a Weyl chamber, and its closure is given by

$$\{t \in L(T);\quad \epsilon_r\theta_r(t) \geq 0 \quad \text{all } r\}.$$

So we can say that the hyperplanes of the diagram divide L(T) into Weyl chambers.

A wall of a Weyl chamber is the intersection of its closure with a hyperplane $L(U_r)$ when the intersection has dimension k - 1.

W permutes the planes of the diagram and the Weyl chambers, by 4.37.

For the following theorem, we suppose chosen an invariant norm in L(G). The word "reflection" is interpreted by using this norm.

5.13 THEOREM

(i) W permutes the Weyl chambers simply transitively.

(ii) For each r, W includes the reflection in the plane $L(U_r)$.

(iii) Such reflections generate W.

(iv) More precisely, for any Weyl chamber B, the

reflections in the walls of B generate W.

(v) Let $p \in L(T)$ and W_p be the stabiliser of p. Then W_p permutes simply transitively the Weyl chambers whose closures contain p.

(vi) W_p is generated by reflections in the planes $L(U_r)$ which contain p.

(vii) More precisely, it suffices to consider those planes which are walls of a fixed Weyl chamber B_o such that $p \in \mathrm{Cl}\, B_o$.

Proof. By taking $p = 0$, we see that (v) $\Rightarrow$ (i), (vi) $\Rightarrow$ (iii) and (vii) $\Rightarrow$ (iv), so we need to prove only (ii), (v), (vi), (vii).

(ii) For each r, W contains an element $\varphi_r \neq 1$ which fixes U_r (5.8), and hence fixes $L(U_r)$, and preserves the inner product in L(T). φ_r can only be the reflection in the plane $L(U_r)$.

(v) Firstly, W_p acts simply. We split the proof into two lemmas:

5.14 **LEMMA.** If $v \in L(T)$ is fixed by some $\psi \in W$, $\psi \neq 1$, then $v \in L(U_r)$ for some r.

Proof. Suppose $n \in N(T)$, $n \notin T$, and n fixes v. Then n fixes the 1-parameter subgroup H corresponding to v (2.17).

Hence there is a maximal torus U containing n and H (4.25). Therefore H lies in two distinct maximal tori, so $H \subset \bigcup U_r$ and $v \in L(U_r)$ for some r.

5.15 LEMMA. If $\psi B = B$ for some Weyl chamber B, $\psi \in W$, them $\psi = 1$.

Proof. W is finite so $\psi^q = 1$ some $q > 0$. Let $v \in B$. Then $v' = \frac{1}{q}\Sigma_1^q \psi^r v$ lies in B and is fixed by ψ. If $\psi \neq 1$, (5.14) shows that v' lies in some $L(U_r)$, which contradicts the hypothesis.

Continuation of 5.13

(v) Secondly, W_p acts transitively on the Weyl chambers whose closures contain p, as follows.

Let B_o, B' be Weyl chambers containing p in their closure, and let $x_o \in B_o$, $x' \in B'$. Since, for $r \neq s$, $L(U_r) \cap L(U_s)$ has dimension $k - 2$ (5.5), there is a polygonal path from x_o to x' not meeting any $L(U_r) \cap L(U_s)$, not meeting any $L(U_r)$ unless it contains p, and meeting each $L(U_r)$ transversely. [Take the path $x_o px'$, and move it slightly.]

Suppose this path crosses $(LU_{k_1}), \ldots, L(U_{k_r})$ successively to get from B_o to B_1 ... to $B_r = B'$. Then $\varphi_{k_r} \ldots \varphi_{k_2} \varphi_{k_1}$

maps B_o via $B_1, \ldots, B_{r-1}$ to $B_r = B'$. Thus W_p is transitive on the Weyl chambers whose closure contains p.

(vi) Let $\psi \in W_p$ and choose B_o such that $p \in \mathrm{Cl}\, B_o$. Set $B' = \psi B_o$. Then $B' = \varphi_{k_r} \ldots \varphi_{k_1} B_o$, with the notation above, so $\psi^{-1}\varphi_{k_r} \ldots \varphi_{k_1} B_o = B_o$. But W_p acts simply, so $\psi^{-1}\varphi_{k_r} \ldots \varphi_{k_1} = 1$ or $\psi = \varphi_{k_r} \ldots \varphi_{k_1}$. Thus these reflections generate W_p.

(vii) Write $\psi = \varphi_{k_r} \ldots \varphi_{k_1}$ as above, and suppose as an inductive hypothesis that we have written $\varphi_{k_s} \ldots \varphi_{k_1}$ as a product of reflections in the walls of B_o. This is trivially possible for s = 1. Then $\varphi_{k_{s+1}}$ is the reflection in a wall $L(U_{k_{s+1}})$ of B_s. But $\varphi_{k_1}^{-1} \ldots \varphi_{k_s}^{-1}$ maps B_s to B_o and $L(U_{k_{s+1}})$ to, say, $L(U_m)$, which contains p. Then

$$\varphi_{k_s} \ldots \varphi_{k_1}\varphi_m\varphi_{k_1}^{-1} \ldots \varphi_{k_s}^{-1} = \varphi_{k_{s+1}} .$$

Therefore

$$\varphi_{k_{s+1}} \ldots \varphi_{k_1} = \varphi_{k_s} \ldots \varphi_{k_1}\varphi_m .$$

Hence ψ may be written as a product of reflections in the walls of B_o which contain p.

5.16 COROLLARY. Divide L(T) into orbits under W. Then each orbit contains precisely one point in the closure of each Weyl chamber B.

Proof. Cl B contains at least one point of each orbit, as

follows. Let $v \in L(T)$. Then $v \in \mathrm{Cl}\, B'$ for some Weyl chamber B', and $B = wB'$ for some $w \in W$. Then $wv \in \mathrm{Cl}\, B$.

Cl B contains not more than one point of each orbit, as follows. Let $p, q \in \mathrm{Cl}\, B$, and $p = wq$. Then $p \in \mathrm{Cl}(wB)$. Since W_p is transitive on those Weyl chambers whose closure contains p, there is $w' \in W_p$ such that $w'wB = B$. Then $w'w = 1$ so $p = w'p = w'wq = q$.

5.17 EXAMPLES

(i) $G = U(n)$. G_e consists of the skew Hermitian matrices. Define an inner product on G_e by $\langle X, Y\rangle = \mathrm{tr}(\bar{X}^T Y) = \mathrm{tr}(-XY)$. This is invariant under G. Restricted to L(T) this has the form (up to a factor $4\pi^2$) $x_1^2 + \ldots + x_n^2$. This is the 'usual' inner product, so reflection is the 'usual' reflection.

The root $\theta_{rs} = x_r - x_s$ gives the plane $x_r = x_s$ for $L(U_{rs})$ and reflection in this plane is given by

$$y_1 = x_1, \ldots, y_r = x_s, \ldots, y_s = x_r, \ldots, y_n = x_n.$$

This is indeed induced by an inner automorphism, namely, by conjugation with

$$
\begin{array}{c|ccccccccccc|}
 & & & & r & & & & s & & & \\
\hline
 & 1 & & & & & & & & & & \\
 & & \ddots & & & & & & & & & \\
 & & & 1 & & & & & & & & \\
r & & & & 0 & & & & 1 & & & \\
 & & & & & 1 & & & & & & \\
 & & & & & & \ddots & & & & & \\
 & & & & & & & 1 & & & & \\
s & & & & -1 & & & & 0 & & & \\
 & & & & & & & & & 1 & & \\
 & & & & & & & & & & \ddots & \\
 & & & & & & & & & & & 1 \\
\hline
\end{array}
$$

[We write -1 and not +1 for the sake of the next example.]

Thus W is the symmetric group on $x_1, \ldots, x_n$. The order $|W|$ of the Weyl group W is $n!$.

(ii) $G = SU(n)$. The same calculation may be repeated, and W is the symmetric group on $x_1, \ldots, x_n$. $|W| = n!$.

(iii) $G = Sp(n)$. This time W consists of transformations of the form

$$y_1 = \epsilon_1 x_{\rho(1)}, \ \ldots, y_n = \epsilon_n x_{\rho(n)},$$

where $\epsilon_r = \pm 1$ each r, and ρ is a permutation. $|W| = n!2^n$.

(iv) $G = SO(2n+1)$. This gives the same Weyl group as $Sp(n)$. $|W| = n!2^n$.

(v) $G = SO(2n)$. W consists of transformations of the form

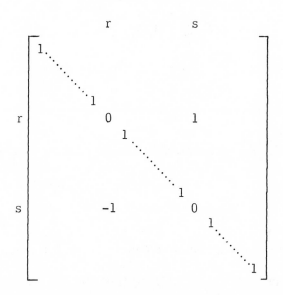

$$y_1 = \epsilon_1 x_{\rho(1)}, \ldots, y_n = \epsilon_n x_{\rho(n)}$$

where $\epsilon_r = \pm 1$ each r, $\prod_1^n \epsilon_r = +1$, and ρ is a permutation. $|W| = n!2^{n-1}$.

5.18 DISCOURSE. The roots θ_r are real linear forms on $L(T)$, that is, they are elements of $L(T)^*$. W acts on $L(T)^*$ by $(wh)(v) = h(w^{-1}v)$.

We have an invariant inner product on $L(T)$, so we may identify $L(T)$ and $L(T)^*$ by $i : L(T) \to L(T)^*$, where we set $(iv_1)(v_2) = \langle v_1, v_2 \rangle$. This commutes with the action of W, so all results on the action of W on $L(T)$ can be transferred to $L(T)^*$ under the isomorphism i. We put an inner product on $L(T)^*$ by copying that of $L(T)$, that is,

$$\langle iv_1, iv_2 \rangle = \langle v_1, v_2 \rangle$$

or, if you prefer,

$$\langle h, iv \rangle = h(v).$$

This is, of course, invariant under W.

φ_r acts on $L(T)$ fixing those vectors v for which $\theta_r(v) = 0$. Therefore φ_r acts on $L(T)^*$ fixing those vectors iv for which $\theta_r(v) = 0$, that is, $\langle \theta_r, iv \rangle = 0$. Thus φ_r fixes those vectors perpendicular to θ_r, and so φ_r is reflection in the plane perpendicular to θ_r.

Note that reflection in the plane perpendicular to the unit vector v is given by $w \to w - 2\langle v, w\rangle v$.

5.19 PROPOSITION

$$\varphi_r(h) = h - \frac{2\langle\theta_r, h\rangle}{\langle\theta_r, \theta_r\rangle}\theta_r.$$

This follows from the discourse.

5.20 DEFINITION. A <u>weight</u> is an element of $L(T)^*$ which takes integer values on the integer lattice.

For example, each root is a weight.

W sends weights to weights. Hence:

5.21 PROPOSITION. If λ is a weight, then

$$\varphi_r(\lambda) = \lambda - \frac{2\langle\theta_r, \lambda\rangle}{\langle\theta_r, \theta_r\rangle}\theta_r$$

is a weight.

W also sends roots to roots, so:

5.22 PROPOSITION. If θ_s is a root, then

$$\varphi_r(\theta_s) = \theta_s - \frac{2\langle\theta_r, \theta_s\rangle}{\langle\theta_r, \theta_r\rangle}\theta_r$$

is a root $\pm\theta_t$.

5.23 **EXAMPLE.** $\varphi_r(\theta_r) = -\theta_r$.

5.24 **PROPOSITION.** In 5.21 and 5.22 the coefficient

$$\frac{-2\langle\theta_r,\lambda\rangle}{\langle\theta_r,\theta_r\rangle}$$

is an integer.

Proof. Choose $v \in L(T)$ so that $\theta_r(v) = 1$. Then $\exp v \in U_r$. φ_r fixes U_r, so $v - \varphi_r(v)$ is in the integer lattice. Therefore $\lambda(v) - \lambda\varphi_r(v)$ is an integer. That is, $\lambda(v) - (\varphi_r\lambda)(v)$ is an integer; since $\varphi_r^{-1} = \varphi_r$, this shows that

$$\lambda(v) - \lambda(v) + \frac{2\langle\theta_r,\lambda\rangle}{\langle\theta_r,\theta_r\rangle}\theta_r(v)$$

is an integer, which is the required result.

5.25 **PROPOSITION.** Let α,β be roots with $\alpha \neq \pm\beta$. Then either

(0) α,β are perpendicular, or

(1) α,β make an angle of 60° or 120° and $|\alpha| = |\beta|$, or

(2) α,β make an angle of 45° or 135° and their ratio is $\sqrt{2}$, or

(3) α,β make an angle of 30° or 150° and their ratio is $\sqrt{3}$.

Proof. We prove this together with:

5.26 PROPOSITION. Let α, β be roots with $\alpha \neq \pm\beta$, and let k be an integer between 0 and $\frac{-2\langle\alpha,\beta\rangle}{\langle\alpha,\alpha\rangle}$ inclusive. Then $\beta + k\alpha$ is also a root.

Proofs. The angle between α and β is given by

$$\cos^2\omega = \frac{\langle\alpha,\beta\rangle^2}{\langle\alpha,\alpha\rangle\langle\beta,\beta\rangle} < 1.$$

Therefore

$$0 \leq \left(\frac{-2\langle\alpha,\beta\rangle^2}{\langle\alpha,\alpha\rangle}\right)\left(\frac{-2\langle\beta,\alpha\rangle}{\langle\beta,\beta\rangle}\right) < 4.$$

By changing the sign of α if necessary, we may suppose $\langle\alpha,\beta\rangle \leq 0$. If $\langle\alpha,\beta\rangle = 0$ then we have case (0) of 5.25, and 5.26 is trivial. Otherwise at least one of

$$\left(\frac{-2\langle\alpha,\beta\rangle}{\langle\alpha,\alpha\rangle}\right), \left(\frac{-2\langle\beta,\alpha\rangle}{\langle\beta,\beta\rangle}\right)$$

is 1. If $\left(\frac{-2\langle\alpha,\beta\rangle}{\langle\alpha,\alpha\rangle}\right) = 1$, then $\beta + \alpha$ is the reflection of β in the plane perpendicular to α, and 5.26 follows in this case. Since 5.25 is symmetric in α and β, we may assume now that $\frac{-2\langle\beta,\alpha\rangle}{\langle\beta,\beta\rangle} = 1$.

Let $\frac{-2\langle\alpha,\beta\rangle}{\langle\alpha,\alpha\rangle} = \nu$, $\nu = 1, 2$, or 3. Then $\frac{\langle\beta,\beta\rangle}{\langle\alpha,\alpha\rangle} = \nu$ so $\frac{|\beta|}{|\alpha|} = \sqrt{\nu}$, and $\cos^2\omega = \frac{\nu}{4}$ so $\cos\omega = \frac{\sqrt{\nu}}{2}$.

If $\underline{\nu = 1}$ we get case (1) of 5.25, and 5.26 has already been demonstrated. We have the following diagram:

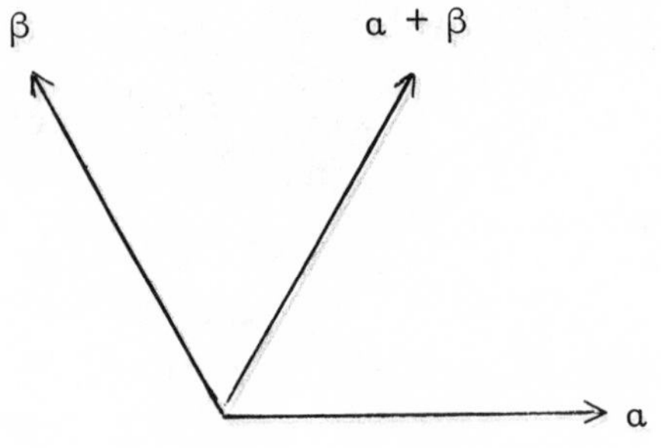

Example. SU(3).

If $\nu = 2$ we get case (2) of 5.25. If we reflect α in the hyperplane perpendicular to β we get $\beta + \alpha$. If we reflect β in the hyperplane perpendicular to α we get $\beta + 2\alpha$, so $\beta + \alpha$, $\beta + 2\alpha$ are roots.

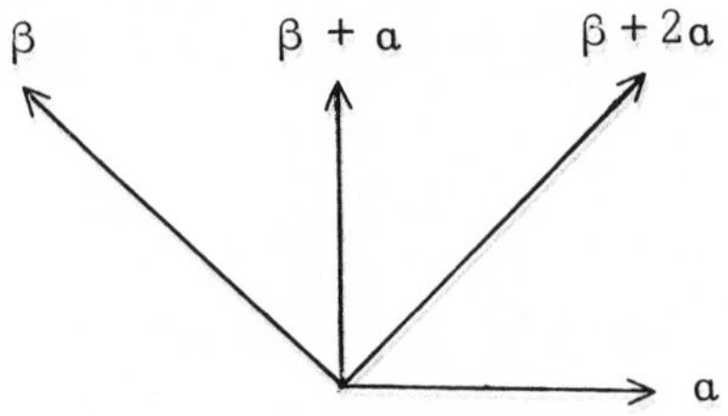

W contains reflections in two hyperplanes at 45°, and so contains the dihedral group D_8.

Examples. Sp(2) or SO(5).

If $\nu = 3$ we get case (3) of 5.25. Reflecting α in the plane perpendicular to β we get $\beta + \alpha$. Reflecting β and $\beta + \alpha$ in the plane perpendicular to α, we get $\beta + 3\alpha$ and $\beta + 2\alpha$.

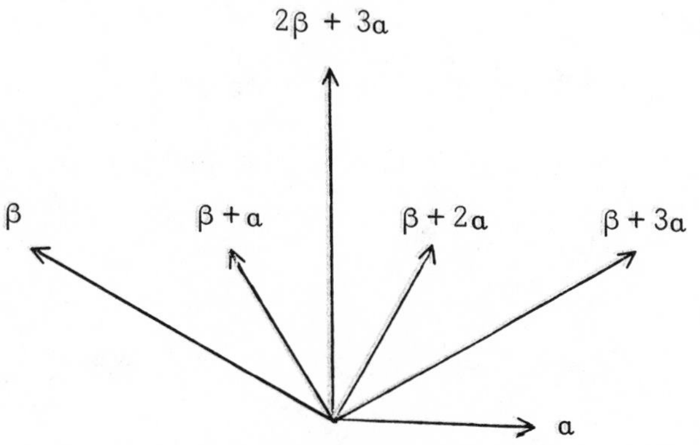

W contains reflections in two hyperplanes at 30°, and so contains the dihedral group D_{12}.

Example. G_2, which is the group of automorphisms of the Cayley numbers as an algebra over R. (It is possible to have fun examining this example explicitly, but we omit this here.)

5.27 **DEFINITION.** Choose a Weyl chamber B in L(T) and call it the fundamental Weyl chamber (FWC). Alter the signs of $\theta_1, \ldots, \theta_m$ so that $\theta_r(v) > 0$ for $v \in B$ all r. Then

$$\{v \in L(T);\quad \theta_r(v) > 0 \quad \text{all } r\} = B.$$

The roots $\theta_1, \ldots, \theta_m$ are now called positive roots, and $-\theta_1, \ldots, -\theta_m$ negative roots.

5.28 **EXAMPLE.** Let G = U(n). Let the fundamental Weyl chamber be given by $x_1 > x_2 > \ldots > x_n$. Then the positive roots are the forms $x_r - x_s$ with $r < s$.

In Sp(n) we take the fundamental Weyl chamber to be given by $x_1 > \ldots > x_n > 0$, and similarly for SO(2n + 1).

For SO(2n) we take $x_1 > x_2 \ldots > x_{n-1} > x_n > -x_{n-1}$.

5.29 **LEMMA.** Let the θ_r be the positive roots and $\lambda_r \geq 0$ in R. Then $\Sigma\lambda_r\theta_r = 0$ implies $\lambda_r = 0$ for all r.

Proof. Take $v \in B$. Then $(\Sigma\lambda_r\theta_r)v = 0$, so $\Sigma\lambda_r(\theta_r v) = 0$, so each λ_r is 0.

5.30 **DEFINITION.** α is a simple root if

(i) α is a positive root, and

(ii) we cannot have $\alpha = \beta + \gamma$ for β and γ positive roots.

5.31 **PROPOSITION.** Any positive root α can be written as a linear combination of simple roots with non-negative integer coefficients.

Proof. If α is not simple, then $\alpha = \beta + \gamma$ where β, γ are positive roots. If either is not simple, we may repeat the process. If this never terminates, since the number of roots is finite, there is an expression $\beta = \beta + \delta_1 + \ldots + \delta_r$ somewhere, contradicting 5.29.

5.32 **LEMMA.** If α, β are distinct simple roots, then $\langle \alpha, \beta \rangle \leq 0$.

Proof. Suppose $\langle \alpha, \beta \rangle > 0$. Then $\frac{2\langle \alpha, \beta \rangle}{\langle \alpha, \alpha \rangle} > 0$ and is an integer, so $\frac{2\langle \alpha, \beta \rangle}{\langle \alpha, \alpha \rangle} \geq 1$. By 5.26, $\beta - \alpha$ is a root. Therefore $\beta - \alpha$ or $\alpha - \beta$ is a positive root, whence either $\beta = (\beta - \alpha) + \alpha$ or $\alpha = (\alpha - \beta) + \beta$ is not simple, contradicting the hypothesis.

5.33 **PROPOSITION.** The simple roots are linearly independent.

Proof. Suppose $v = \Sigma \mu_r \theta_r = \Sigma \nu_s \theta_s$ where the θ_r are simple roots, all μ_r, ν_s are non-negative, and the sums run over disjoint sets of subscripts. Then

$$\langle v, v \rangle = \Sigma \mu_r \nu_s \langle \theta_r, \theta_s \rangle \leq 0.$$

Therefore $v = 0$ and we may apply 5.29.

5.34 **COROLLARY.** The fundamental Weyl chamber is given by

$$\theta_1(v) > 0, \ldots, \theta_s(v) > 0,$$

where $\theta_1, \ldots, \theta_s$ are the simple roots.

Proof. This is clear from 5.31.

So simple roots correspond to walls of the fundamental Weyl chamber.

5.35 EXAMPLE. $G = U(n)$. The fundamental Weyl chamber is $x_1 > x_2 \ldots > x_n$. The simple roots are

$$x_1 - x_2, x_2 - x_3, \ldots, x_{n-1} - x_n.$$

Any other root can be written as a linear sum of these, e.g.,

$$x_r - x_s = (x_r - x_{r-1}) + \ldots + (x_{s-1} - x_s)$$

for $r < s$. And

$$x_1 - x_2, \ldots, x_{n-1} - x_n$$

are linearly independent.

5.36 EXERCISE. If α is a simple root and we write $\alpha = \Sigma\mu_r\theta_r$, where the μ_r are non-negative numbers and the θ_r are positive roots, then we have written $\alpha = \alpha$.

5.37 DEFINITION. The Dynkin diagram is constructed as follows. Take one node for each simple root α. Given two distinct simple roots α, β join the corresponding nodes by $\nu = 0, 1, 2$ or 3 bonds, where ν follows 5.26.

5.38 EXAMPLE. $G = U(n)$. Between $(x_r - x_{r+1})$ and $(x_{r+1} - x_{r+2})$, $\nu = 1$. Otherwise $\nu = 0$. Hence the Dynkin diagram is 0——0——0 ... 0——0 with $n - 1$ nodes.

5.39 LEMMA. If θ_r is a simple root then φ_r permutes the positive roots except θ_r, which goes to $-\theta_r$.

Proof. We give two proofs.

(i) Choose a point v of the diagram such that $\theta_r(v) = 0$ and $\theta_s(v) > 0$ for any other simple root θ_s. Then $\theta_t(v) > 0$ for any positive root θ_t other than θ_r.

Let S be a spherical neighbourhood of v not meeting any plane $\theta_t = 0$ for $t \neq r$. Let $w \in S \cap$ (FWC). Then $\varphi_r(w) \in S$. Therefore

$$(\varphi_r\theta_t)(w) = \theta_t(\varphi_r w) > 0$$

for $t \neq r$. Thus $\varphi_r\theta_t$ is a positive root.

(ii) Let $\theta_1, \ldots, \theta_s$ be the simple roots, and let θ_t be a positive root. Write

$$\theta_t = n_1\theta_1 + \ldots + n_s\theta_s.$$

Then

$$\varphi_r(\theta_t) = \theta_t - \frac{2\langle\theta_r, \theta_t\rangle}{\langle\theta_r, \theta_r\rangle}\theta_r$$

differs from θ_t only in the coefficient of θ_r. Therefore $\varphi_r(\theta_t)$ has at least one positive coefficient if $\theta_t \neq \theta_r$ and so (5.31 and 5.33) $\varphi_r(\theta_t)$ is a positive root.

5.40 DEFINITION. The fundamental dual Weyl chamber (FDWC) is the set of points in L(T)* corresponding under i to the fundamental Weyl chamber in L(T). That is, the FDWC is the set of $h \in L(T)^*$ such that $\langle \theta_r, h \rangle > 0$ for each simple root θ_r.

5.41 DEFINITION. Let $\theta_1, \ldots, \theta_m$ be the positive roots. Define $\beta \in L(T)^*$ by $\beta = \frac{1}{2}(\theta_1 + \ldots + \theta_m)$. This is not necessarily a weight.

5.42 PROPOSITION. β lies in the fundamental dual Weyl chamber. Indeed, $\frac{2\langle \alpha, \beta \rangle}{\langle \alpha, \alpha \rangle} = 1$ for each simple root α.

Proof. Let $\alpha = \theta_r$. Then φ_r permutes the positive roots other than α. There are three cases:

(i) $\varphi_r(\theta_t) = \theta_t$. Then $\langle \theta_r, \theta_t \rangle = 0$ so θ_t contributes 0 to $\langle \alpha, \beta \rangle$.

(ii) θ_r permutes θ_t and θ_u, $t \neq u$. Then

$$\langle \theta_r, \theta_t + \theta_u \rangle = 0,$$

so $\theta_t + \theta_u$ contributes 0.

(iii) $\theta_t = \theta_r$. This case contributes $\frac{1}{2}\langle \alpha, \alpha \rangle$ to $\langle \alpha, \beta \rangle$.

Therefore $\frac{2\langle \alpha, \beta \rangle}{\langle \alpha, \alpha \rangle} = 1$.

5.43 EXERCISES. Work out $\frac{1}{2}(\theta_1 + \ldots + \theta_m)$ for the following groups:

(i) SU(3).

(ii) SO(5).

(iii) G_2.

5.44 PROPOSITION. In L(T), reflections in the planes $\theta_r = k$ for $k \in Z$ cover the action of φ_r on T.

Proof. Let $v \in L(T)$ be such that $\theta_r(v) = k$. Then the reflection is given by

$$x \to \varphi_r(x - v) + v = \varphi_r(x) - \varphi_r(v) + v.$$

But v maps into U_r, so $\varphi_r(v)$ and v have the same image in T. Therefore $\varphi_r(x)$ and $\varphi_r(x) - \varphi_r(v) + v$ have the same image in T.

5.45 DEFINITION. The extended Weyl group Γ is the group generated by reflections in all the planes $\theta_r = k$, $k \in Z$, of the diagram.

By 5.44, Γ covers the action of W on T. Define $\Gamma_o = \mathrm{Ker}\,(\Gamma \to W)$.

5.46 DISCOURSE. We have a split extension

$$\begin{array}{ccccc} \Gamma_o & \longrightarrow & \Gamma & \longrightarrow & W \\ & & \uparrow & \nearrow_{1} & \\ & & W & & \end{array}$$

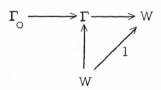

Γ_o is the subgroup of trans ations. Each one is the translation by an element of the in eger lattice I, so we can regard Γ_o as a subgroup of I. (It is not necessarily the whole of I.)

Our next object is to calculate the fundamental group $\pi_1(G)$ in terms of the Stiefel diagram. The topological invariant $\pi_1(G)$ may be distasteful to some algebraists, and so some remarks are in order about the use to be made of it. First, one of the main theorems (6.41) is classically stated with the condition "$\pi_1(G) = 0$", and some of the subsidiary results used in its proof use the same condition. However, we are just going to prove (5.47) "$\pi_1(G) \cong I/\Gamma_o$", so it would be possible to rewrite 6.41 with the data in the form "$\Gamma_o = I$", which after all is what is used in the proof of 6.41. Secondly, we propose to use $\pi_1(G)$ to classify the connected covering groups over G, as is usual in algebraic topology. For our arguments to proceed without this (notably at 5.56 below) it would be necessary to construct the double covering Spin(n) of SO(n) without reference to π_1; and of course this is possible by pure algebra, for example, using Clifford algebras. This is an interesting chapter of algebra, but it involves more work without providing so much more insight. Sometimes one can buy algebraic purity at too high a price [23].

To continue: we have $I \cong \pi_1(T)$, as follows. Consider $I \subset L(T) \to T$. For $v \in I$ choose a path ω in $L(T)$ from some $\omega(o)$ to $\omega(1) = v + \omega(0)$. Its projection is a closed path in T, and so represents an element of $\pi_1(T)$, since $\pi_1(T)$ is Abelian.

The map $i : T \to G$ induces $I = \pi_1(T) \xrightarrow{i_*} \pi_1(G)$.

5.47 **THEOREM.** i_* is epi and induces $I/\Gamma_o \cong \pi_1(G)$.

Proof. 5.48-5.55 will, together, form a proof.

5.48 **PROPOSITION.** Let γ_r be the reflection of 0 in the plane $\theta_r = 1$. Then Γ_o is the subgroup of I generated by the γ_r.

Proof. Γ_o contains each γ_r, since reflection in $\theta_r = 0$ followed by reflection by $\theta_r = 1$ is translation by γ_r.

Conversely, we claim that, if $\gamma \in \Gamma$, then $\gamma(0) = \Sigma n_r \gamma_r$, whence, if $\gamma \in \Gamma_o$, γ is translation by $\Sigma n_r \gamma_r$. We prove this claim by induction on the number of reflections used to build up γ.

Suppose $\gamma = \rho\delta$ where ρ is reflection in $\theta_r = k$, and suppose $\delta(0) = \Sigma n_s \gamma_s$. Now

$$\rho(x) = x + (k - \theta_r(x))\gamma_r.$$

Therefore

$$\rho\delta(0) = \Sigma n_s\gamma_s + k\gamma_r - \theta_r(\Sigma n_s\gamma_s)\gamma_r .$$

But $\theta_r(\Sigma n_s\gamma_s)$ is an integer, since $\Sigma n_s\gamma_s$ is in the integer lattice. Therefore $\rho\delta(0)$ has the required form.

5.49 EXAMPLES.

(i) $G = U(n)$ or $SU(n)$. The reflection of 0 in $x_r - x_s = 1\,(r < s)$ is the point $(0 \ldots 0\ \overset{r}{1}\ 0 \ldots 0\ \overset{s}{-1}\ 0 \ldots 0)$. Define $\pi : I \to Z$ by $\pi(x_1, \ldots, x_n) = x_1 + \ldots + x_n$. Then $\Gamma_o = \operatorname{Ker} \pi$. For $SU(n)$ we have $I/\Gamma_o = 0$. For $U(n)$ we have $I/\Gamma_o \cong Z$.

(ii) $G = Sp(n)_r$. The reflection of 0 in $2x_r = 1$ is $(0 \ldots 0\ \overset{r}{1}\ 0 \ldots 0)$. We have $I/\Gamma_o = 0$.

(iii) $G = SO(2n)$ or $SO(2n + 1)$. The reflection of 0 in $x_r - x_s\,(r < s)$ is $(0 \ldots 0\ \overset{r}{1}\ 0 \ldots 0\ \overset{s}{-1}\ 0 \ldots 0)$. The reflection of 0 in $x_r + x_s = 1$ is $(0 \ldots 0\ \overset{r}{1}\ 0 \ldots 0\ \overset{s}{1}\ 0 \ldots 0)$. For $SO(2n + 1)$ the reflection of 0 in $x_r = 1$ is $(0 \ldots 0\ \overset{r}{2}\ 0 \ldots 0)$, which gives nothing new. Define $\pi : I \to Z_2$ by $\pi(x_1, \ldots, x_n) = x_1 + \ldots + x_n \bmod 2$. Then $\Gamma_o = \operatorname{Ker} \pi$. Thus $I/\Gamma_o \cong Z_2$.

In the special case of $SO(2)$, $\Gamma_o = 0$ and $I/\Gamma_o \cong Z$.

5.50 LEMMA. $I \cong \pi_1(T) \to \pi_1(G)$ maps Γ_0 to 0.

Proof. We show that γ_r goes to zero. Well, let ω be a rectilinear path from 0 to γ_r in L(T). Then
$\exp \omega(1 - t) = \varphi_r \exp \omega(t)$ for $0 \le t \le \frac{1}{2}$. By 5.9, we can find
$g \in G$ such that $\varphi_r(x) = gxg^{-1}$, so that
$\exp \omega(1 - t) = g \exp \omega(t)g^{-1}$, and such that there is a path from g to e each point of which keeps U_r fixed. So
$\exp \omega(1 - t)$ is homotopic to $e \exp \omega(t)\ e^{-1} = \exp \omega(t)$, keeping
$t = 0$, $t = \frac{1}{2}$ fixed. Hence $\exp \omega(t)$ for $0 \le t \le 1$ is contractible keeping end points fixed. So γ_r goes to zero in $\pi_1(G)$.

5.51 NOTATION. Let $G_R, T_R, L(T)_R$ denote the sets of regular points in G, T, L(T) respectively.

5.52 LEMMA. $i_* : \pi_1(G_R) \to \pi_1(G)$ is an isomorphism.

Proof. The complement of G_R has Hausdorff dimension $\le n - 3$, by 4.42 and standard Hausdorff dimension theory, and the result follows by standard homotopy theory.

5.53 LEMMA. Define $f_R : G/T \times T_R \to G_R$ by $f_R(g,t) = gtg^{-1}$. Then f_R is a covering with fibre W.

Proof. W acts on the left on G/T as follows. Let $\varphi \in W$ and let $n \in N(T)$ represent φ. Define

$$\varphi(gT) = gTn^{-1} = gn^{-1}T.$$

W also acts on the left on T_R, and so acts on $G/T \times T_R$. Let $G/T\times_W T_R$ be the orbit space. Since W acts freely on G/T, the projection

$$G/T \times T_R \to G/T\times_W T_R$$

is a covering with fibre W.

Now f_R factors through $G/T\times_W T_R$, and $G/T\times_W T_R \to G_R$ is a one-one and onto map between manifolds of the same dimension, and so is a homeomorphism. Hence the result.

5.54 LEMMA. $i_* : \pi_1(T) \to \pi_1(G)$ is epi.

Proof. Consider the map

$$G/T \times T_R \xrightarrow{f_R} G_R \subset G,$$

where f_R is a finite cover. Let the components of T_R be T_R^i; then since G/T is connected, the components of $G/T \times T_R$ are $G/T \times T_R^i$; and so each of the following maps is monomorphic.

$$\pi_1(G/T \times pt) \to \pi_1(G/T \times T_R^i) \xrightarrow{f_{R*}} \pi_1(G_R) \to \pi_1(G).$$

Now the map $G/T \times t_o \to G$, given by $g \to gt_o g^{-1}$, is nullhomotopic by taking a path from t_o to e. So $\pi_1(G/T) = 0$.

Hence, from the homotopy exact sequence of a fibration

$$\pi_1(T) \to \pi_1(G) \to \pi_1(G/T),$$

we deduce that $\pi_1(T) \to \pi_1(G)$ is epi.

5.55 LEMMA. If $v \in I$ maps to 0 under $I \cong \pi_1(T) \to \pi_1(G)$, then $v \in \Gamma_o$.

Proof. We may suppose that, for any $\gamma \in \Gamma_o$, $v + \gamma$ is not closer than v to the origin in I. Then $\theta_r(v) = -1$, 0 or 1 for each root θ_r, for, if $\theta_r(v) > 1$, then the reflection of v in $\theta_r = 1$ is closer to the origin, and correspondingly if $\theta_r(v) < -1$.

Let ω be the linear path in L(T) from $\omega(0) = 0$ to $\omega(1) = v$. This does not cross any diagram planes, although it may lie in some, and may meet others at $\omega(0)$ and $\omega(1)$. So there is a linear path ω' from $\omega'(0)$ to $\omega'(1) = \omega'(0) + v$ which is close to ω and which meets diagrams planes only close to $\omega'(1)$.

Consider the diagram

$$\begin{array}{ccc} G/T \times L(T)_R & \xrightarrow{f_R} & G_R \\ \cap & & \cap \\ G/T \times L(T) & \xrightarrow{f} & G \end{array}.$$

By taking the identity coset in G/T, the path ω' may be considered as in $G/T \times L(T)$. Then $f\omega'$ is a loop in G which lies in G_R except near $f\omega'(1)$. By 4.42, we may move this loop slightly near $f\omega'(1)$ so that it lies in G_R, and this loop is contractible in G_R. Since $G/T \times L(T)_R \to G_R$ is a covering, we may now lift the loop to a path ω'' in $G/T \times L(T)_R$ starting near $T \times 0$. Then ω'' will be the same as ω' except near $\omega'(1)$. Further, since we have altered $f\omega'$ only near e in G, the projection of ω'' onto the factor $L(T)$ is close to ω'. Now $f_R\omega''$ is contractible in G_R, so ω'' is a closed loop in $L(T)_R$, and v is approximately zero. But v is in I, so $v = 0$.

5.56 **DISCUSSION.** We have now shown (5.47 and 5.49) that $\pi_1(SO(m)) \cong Z_2$ for $m > 2$. Therefore SO(m) has a double cover called Spin(m). It is clear that the cover of a maximal torus in SO(m) is a maximal torus in Spin(m). Take as the standard maximal torus $\tilde{T}$ in Spin(m) the cover of the standard maximal torus T in SO(m). Then $L(\tilde{T}) \cong L(T)$ under the covering map, though this does not preserve the integer lattices. I consists of all $(x_1, \ldots, x_n)$ with all x_r integers, and $\tilde{I}$ consists of all $(x_1, \ldots, x_n)$ with all x_r integers and $x_1 + \ldots + x_n$ even. Similarly $L(\tilde{T})^* \cong L(T)^*$, but this does not preserve the lattices of weights. For example, $\frac{1}{2}(x_1 + \ldots + x_n)$ is not a weight in SO(m) but is one in Spin(m).

Now Ad : $G \to SO(n)$ induces

$Ad_* : \pi_1(G) \to \pi_1(SO(n)) \cong Z_2$ (for $n > 2$).

We distinguish two cases.

(i) Ad_* is zero, and we can lift Ad to get the following diagram.

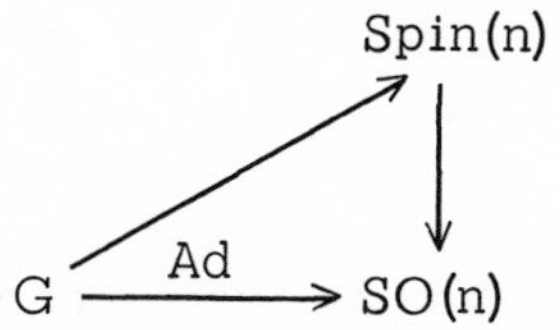

(ii) Ad_* is non-zero. Then Ad defines a double cover $\tilde{G}$ of G, and we have the following diagram.

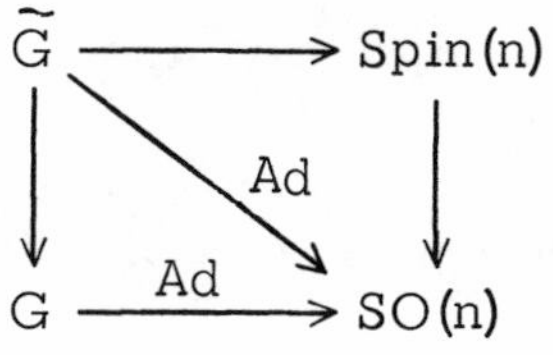

For $\tilde{G}$, (i) applies. By 3.68, the representation theory of $\tilde{G}$ determines that of G. So, in what follows, we will assume that (i) applies.

5.57 PROPOSITION. In this case, $\beta = \frac{1}{2}(\theta_1 + \ldots + \theta_m)$ (see 5.41) is a weight.

<u>Proof</u>. In 4.12 we split G_e as a T-space in the form

$V_o \oplus \Sigma_1^m V_i$. Choose bases for $V_1, \ldots, V_m, V_o$, and put them together in this order to form a base for G_e. Then the composition $T \subset G \xrightarrow{Ad} \text{Aut } G_e = SO(n)$ sends T into the standard maximal torus T' of $SO(n)$. Further, if $L(T') \xrightarrow{x_r} R$ denotes the rth co-ordinate function, then the composition $L(T) \to L(T') \xrightarrow{x_r} R$ is the root $\pm\theta_r$, for $r \leq m$, or zero, for $r > m$. With the same sign attached to each θ_r, we now have $\pm\theta_1 \pm \ldots \pm \theta_m = (\Sigma x_r)Ad$.

Now Ad lifts to Spin(n), and $\frac{1}{2}\Sigma x_r$ is a weight for Spin(n), so $(\frac{1}{2}\Sigma x_r)Ad$ is a weight for G. Thus $\frac{1}{2}(\pm\theta_1 \ldots \pm\theta_m)$ is a weight for G, and so is $\beta = \frac{1}{2}(\theta_1 + \ldots + \theta_m)$, as this differs from $\frac{1}{2}(\pm\theta_1 \ldots \pm\theta_m)$ by a sum of positive roots.

5.58 **LEMMA.** In this case $\omega \to \omega + \beta$ gives a one-one correspondence between weights $\omega \in$ Cl FDWC and weights $\omega + \beta \in$ FDWC.

Proof

(i) If ω is a weight and $\langle \omega, \theta_r \rangle \geq 0$ for all simple roots θ_r then $\langle \omega + \beta, \theta_r \rangle > 0$ by 5.42.

(ii) If ω is a weight and $\langle \omega, \theta_r \rangle > 0$ for all simple roots θ_r then $\frac{2\langle \omega, \theta_r \rangle}{\langle \theta_r, \theta_r \rangle} > 0$ and is an integer (5.24), so ≥ 1. Now

$\dfrac{2\langle\beta,\theta_r\rangle}{\langle\theta_r,\theta_r\rangle} = 1$, so $\dfrac{2\langle\omega-\beta,\theta_r\rangle}{\langle\theta_r,\theta_r\rangle} \geq 0$ and $\omega - \beta$ is a weight in Cl FDWC.

We showed (5.24) that, if ω is a weight and θ_r a root, then $\dfrac{2\langle\theta_r,\omega\rangle}{\langle\theta_r,\theta_r\rangle}$ is an integer. We now examine the converse.

5.59 PROPOSITION. If $\dfrac{2\langle\theta_r,\omega\rangle}{\langle\theta_r,\theta_r\rangle}$ is an integer for some $\omega \in L(T)^*$ and all simple roots θ_r, then it is an integer for all roots θ_r.

Proof. Suppose $\dfrac{2\langle\theta_r,\omega\rangle}{\langle\theta_r,\theta_r\rangle}$ is an integer for all simple roots θ_r and also for the root θ_s. Let φ_r correspond to some simple root θ_r, and let $\theta_t = \varphi_r(\theta_s)$. Then $\langle\theta_t,\theta_t\rangle = \langle\theta_s,\theta_s\rangle$ and so

$$\frac{2\langle\theta_t,\omega\rangle}{\langle\theta_t,\theta_t\rangle} = \frac{2}{\langle\theta_s,\theta_s\rangle}\left\langle\theta_s - \frac{2\langle\theta_r,\theta_s\rangle}{\langle\theta_r,\theta_r\rangle}\theta_r\,,\ \omega\right\rangle$$

$$= \frac{2\langle\theta_s,\omega\rangle}{\langle\theta_s,\theta_s\rangle} - \frac{2\langle\theta_r,\omega\rangle}{\langle\theta_r,\theta_r\rangle}\cdot\frac{2\langle\theta_r,\theta_s\rangle}{\langle\theta_s,\theta_s\rangle}$$

which is an integer.

But the reflections φ_r generate W (5.34 and 5.13(iv)) and any root θ_s can be written as $\varphi\theta_r$ for some simple root θ_r and some $\varphi \in W$, by considering θ_s as the wall of a Weyl chamber and throwing this chamber onto the FWC (5.34).

Hence the result.

5.60 PROPOSITION. Suppose $\frac{2\langle\theta_r,\omega\rangle}{\langle\theta_r,\theta_r\rangle}$ is an integer for some $\omega \in L(T)^*$ and each simple root θ_r. Then ω takes integer values on Γ_o.

Proof. Γ_o is generated by the points γ_r, where $\gamma_r = v - \varphi_r v$ for any v such that $\theta_r v = 1$. We have

$$\begin{aligned}\omega(\gamma_r) &= \omega v - \omega(\varphi_r v)\\ &= (\omega - \varphi_r\omega)(v)\\ &= \frac{2\langle\theta_r,\omega\rangle}{\langle\theta_r,\theta_r\rangle}\theta_r(v)\end{aligned}$$

which is an integer. Thus ω is integral on each γ_r and so on Γ_o.

5.61 COROLLARY. If G is simply connected and $\frac{2\langle\theta_r,\omega\rangle}{\langle\theta_r,\theta_r\rangle}$ is an integer for each simple root θ_r, then ω is a weight.

Proof. $\Gamma_o = I$.

5.62 THEOREM. If G is simply connected it has just $k = \dim T$ simple roots $\theta_1,\ldots,\theta_k$, and has weights $\omega_1,\ldots,\omega_k$ such that $\frac{2\langle\theta_r,\omega_t\rangle}{\langle\theta_r,\theta_r\rangle} = \delta_{rt}$. The weights are then the linear combinations $n_1\omega_1 + \ldots + n_k\omega_k$ with $n_r \in Z$ each r. The FDWC consists of all points $\Sigma n_r \omega_r$ with each $n_r > 0$, and the Cl FDWC consists of all points $\Sigma n_r \omega_r$ with each $n_r \geq 0$. Also

$$\frac{1}{2}(\theta_1 + \ldots + \theta_m) = \omega_1 + \ldots + \omega_k.$$

Proof. Suppose there are just $k - \nu$ simple roots. Then all the roots lie in a subspace of $L(T)^*$ of dimension $k - \nu$, and so Γ_o lies in a subspace of $L(T)$ of dimension $k - \nu$. Then I/Γ_o has rank at least ν which implies $\nu = 0$.

There are elements ω_t in $L(T)^*$ such that $\frac{2\langle\theta_r, \omega_t\rangle}{\langle\theta_r, \theta_r\rangle} = \delta_{rt}$, and they are weights by 5.61. Every element ω of $L(T)^*$ can be written $\omega = \Sigma n_r \omega_r$, and then $\frac{2\langle\theta_r, \omega\rangle}{\langle\theta_r, \theta_r\rangle} = n_r$, so ω is a weight if and only if each n_r is an integer. The statements about FDWC follow from the definition (5.40).

Set $\beta = \frac{1}{2}(\theta_1 + \ldots + \theta_m)$. Then $\frac{2\langle\theta_r, \beta\rangle}{\langle\theta_r, \theta_r\rangle} = 1$ (5.42), so $\beta = \omega_1 + \ldots + \omega_k$.

5.63 EXAMPLE. Let $G = SU(n)$.

Take $\omega_t = x_1 + \ldots + x_t$ for $1 \leq t \leq n - 1$. Then

$$\langle x_r - x_{r+1}, \omega_t\rangle = \delta_{rt}$$

so

$$\frac{2\langle\theta_r, \omega_t\rangle}{\langle\theta_r, \theta_r\rangle} = \delta_{rt}.$$

The elements of $L(T)^*$ can be written

$$a_1x_1 + \ldots + a_{n-1}x_{n-1},$$

since $\Sigma x_i = 0$. They lie in FDWC if

$$a_1 > a_2 \ldots > a_{n-1} > 0.$$

Thus they may be written

$$b_1\omega_1 + \ldots + b_{n-1}\omega_{n-1},$$

and lie in FDWC if $b_1, \ldots, b_{n-1} > 0$.

5.64 COUNTEREXAMPLES

(i) Let $G = U(2)$, where $\pi_1(U(2)) \cong Z$. We have $\dim T = 2$, but there is only one root. The FDWC is a half-plane, which cannot be expressed in the given form.

(ii) Let $G = SO(4)$, where $\pi_1(SO(4)) \cong Z_2$. The dual diagram is as follows:

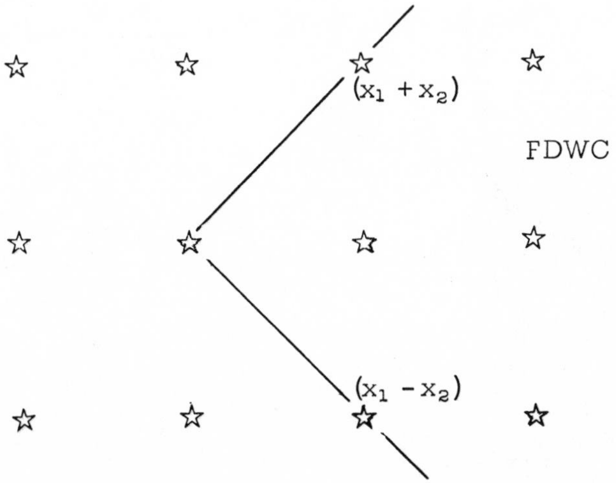

Here the asterisks represent weights. The FDWC is the quarter-plane shown. The weights in the Cl FDWC do not form a free Abelian semi-group, and

$$\omega_1 = \frac{1}{2}(x_1 + x_2), \quad \omega_2 = \frac{1}{2}(x_1 - x_2)$$

are not weights.

Chapter 6

REPRESENTATION THEORY

Notice. Throughout this chapter G is a compact connected Lie group, and T is a maximal torus of G.

6.1 **THEOREM.** (Weyl Integration Formula.) There is a real function u on T such that

$$\int_G f(g) = \int_T f(t)u(t)$$

for all class functions f on G.

Indeed $u(t) = \delta\bar{\delta}/|W|$, where

$$\delta = \Pi_{j=1}^{m}\left(e^{\pi i\theta_j(t)} - e^{-\pi i\theta_j(t)}\right)$$

and θ_j runs over the distinct roots of G.

Proof. Define $f : G/T \times T \to G$ by $f(g,t) = gtg^{-1}$. (See 5.53.) Then f factors through $G/T\times_W T$:

$$\begin{array}{ccc} G/_T \times T & \xrightarrow{\quad b \quad} & G/_T \times_W T \\ & \overset{f}{\searrow} & \Big\downarrow c \\ & & G \end{array}$$

Now c has degree 1, since it is a homeomorphism when restricted to $G/T \times_W T_R \to G_R$, and b is a $|W|$-fold covering. So f has degree $|W|$, and

$$|W| \int_G f \, dg = \int_{G/T \times T} f^* \, dg^*,$$

where f^*, dg^* are the induced function and measure on $G/T \times T$. If f is a class function, then f^* is constant along G/T.

Now we must evaluate det f' at a general point (g,t) of $G/T \times T$. First, let u run through a neighbourhood of e in T. Then

$$f(g,tu) = gtug^{-1} = gtg^{-1}gug^{-1}.$$

Therefore $f'(g,t) = \mathrm{Ad}\, g$, where we consider the first factor fixed. Second, let v be in a transversal V of T in G. Then

$$f(gv,t) = gvtv^{-1}g^{-1},$$

so

$$\begin{aligned} f'(g,t)(dv) &= g(dv)tg^{-1} - gt(dv)g^{-1} \\ &= (gtg^{-1})(gt^{-1}dvtg^{-1} - gdvg^{-1}), \end{aligned}$$

so

$$f'(g,t) = \mathrm{Ad}g(\mathrm{Ad}t^{-1} - I),$$

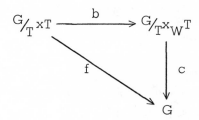

where we consider the second factor fixed. Thus

$$\det f'(g,t) = \det(\mathrm{Ad}\,t^{-1} - I).$$

Now Adt has the form

$$\begin{bmatrix} \cos 2\pi\theta_1 & \sin 2\pi\theta_1 & \\ -\sin 2\pi\theta_1 & \cos 2\pi\theta_1 & \\ & & \ddots \end{bmatrix}$$

so $\mathrm{Ad}\,t^{-1} - I$ has the form

$$\begin{bmatrix} \cos 2\pi\theta_1 - 1 & \sin 2\pi\theta_1 & \\ \sin 2\pi\theta_1 & \cos 2\pi\theta_1 - 1 & \\ & & \ddots \end{bmatrix}$$

and

$$\det(\mathrm{Ad}\,t^{-1} - I) = \prod_1^m \left(\cos^2 2\pi\theta_r - 2\cos 2\pi\theta_r + 1 + \sin^2\theta_r\right)$$
$$= \prod_1^m (4\sin^2 \pi\theta_r) = \prod\left(e^{\pi i\theta_r} - e^{-\pi i\theta_r}\right)\prod\left(e^{-\pi i\theta_r} - e^{\pi i\theta_r}\right)$$
$$= \delta\bar{\delta},$$

where

$$\delta = \prod\left(e^{\pi i\theta_r} - e^{-\pi i\theta_r}\right).$$

Hence the result.

6.2 DEFINITION. W acts on L(T). For $\varphi \in W$, let sign φ denote the sign of the determinant. Then we say that

$\chi \in K(T)$ is a <u>symmetric</u> character if $\varphi\chi = \chi$ for each $\varphi \in W$, and is an <u>anti-symmetric</u> or <u>alternating</u> character if $\varphi\chi = (\text{sign}\,\varphi)\,\chi$.

6.3 **EXAMPLE.** Suppose Ad : $G \to SO(n)$ lifts to Spin(n). Then

$$\delta = \Pi_1^m\left(e^{\pi i\theta_r} - e^{-\pi i\theta_r}\right)$$

is an anti-symmetric character.

<u>Proof</u>.

$$\delta = \Sigma\epsilon_1 \ldots \epsilon_m \operatorname{Exp} \pi i(\epsilon_1\theta_1 + \ldots + \epsilon_m\theta_m)$$

where $\epsilon_i = \pm 1$ and there are 2^m terms. Note that, by 5.57,

$$\frac{1}{2}(\epsilon_1\theta_1 + \ldots + \epsilon_m\theta_m)$$

is a weight, so $\delta \in K(T)$.

Let $x \in N(T)$ represent $\varphi \in W$. Then the action of φ is given by $g \to xgx^{-1}$. This induces a map $G_e \to G_e$ which maps T_e to T_e. On T_e it preserves or reverses orientation according to (sign φ). Also φ permutes $V_1, \ldots, V_m$ (5.5 and 3.22). If φ maps V_j to V_k preserving orientation, then it sends θ_j to θ_k; and if reversing orientation, then it sends θ_j to $-\theta_k$. If it reverses orientation ν times then $\varphi\delta = (-1)^\nu\delta$.

But φ preserves the orientation of G_e, since x may be connected to e by a path. Therefore

$$(\operatorname{sign} \varphi)(-1)^{\nu} = +1.$$

That is,

$$\varphi\delta = (\operatorname{sign} \varphi)\delta.$$

6.4 PROPOSITION. If a character χ (of T) vanishes on U_r then it can be written

$$\chi = [\operatorname{Exp}(2\pi i\theta_r) - 1]\psi,$$

where ψ is a character.

Proof

(i) Suppose U_r has just one component. Then we may take a basis $e_1, \ldots, e_k$ of the integer lattice of L(T) as follows. Let $e_2, \ldots, e_k$ be a basis of the integer lattice of $L(U_r)$, and let e_1 be a point of the integer lattice of L(T) for which $\theta_r(e_1) = 1$. Let $\xi_1, \ldots, \xi_k$ be the characters of the basic representations of T. Then $\operatorname{Exp}(2\pi i\theta_r) = \xi_1$, and we can write $\chi = \Sigma_n c_n \xi_1^n$, where each c_n is a finite Laurent series in $\xi_2, \ldots, \xi_k$.

On U_r, $\xi_1 = 1$ so $\Sigma c_n = 0$. The monomials $\xi_2, \ldots, \xi_k$ are linearly independent on U_r, so $\Sigma_n c_n$ is the zero Laurent series. Set

$$\psi = \Sigma_n(\ldots + c_{n+1} + c_n)\xi_1^{n-1}.$$

This is a finite Laurent series and $\chi = (\xi_1 - 1)\psi$.

(ii) Suppose U_r has two components. Take a basis as before, but with $\theta_r(e_1) = 2$. Then

$$\mathrm{Exp}(2\pi i\theta_r) = \xi_1^2.$$

Consider $\chi = \Sigma c_n \xi_1^n$ and note that U_r is given by $\xi_1 = 1$ and $\xi_1 = -1$. Therefore $\Sigma c_n = 0$ and $\Sigma(-1)^n c_n = 0$. Thus $\Sigma_{n \text{ odd}} c_n = 0$ and $\Sigma_{n \text{ even}} c_n = 0$, and we may argue as before to get $\chi = (\xi_1^2 - 1)\psi$.

6.5 **PROPOSITION.** If χ is an anti-symmetric character, then

$$\chi = \Pi_{j=1}^m [\mathrm{Exp}(2\pi i\theta_j) - 1]\psi,$$

where ψ is a character.

Proof. It is only necessary to show that, if

$$\chi = [\mathrm{Exp}(2\pi i\theta_i) - 1]\psi$$

and χ vanishes on U_r for $r \neq i$, then ψ vanishes on U_r, for we may then argue by induction using 6.4. Well, ψ does indeed vanish on U_r except possibly on $U_i \cap U_r$. But $\dim U_r = k - 1$ and $\dim U_i \cap U_r = k - 2$. Therefore ψ vanishes on all of U_r by continuity.

6.6 THEOREM. Suppose Ad lifts to Spin(n). Then $\psi \to \psi\delta$ gives an isomorphism from the additive group of symmetric characters to the additive group of anti-symmetric characters.

<u>Proof</u>

(i) δ is anti-symmetric (6.3), so the map goes where the theorem says.

(ii) $\frac{1}{|W|}\int \delta\bar{\delta} = 1$, so $\delta \neq 0$ and the map is mono (3.77).

(iii) Suppose χ is an anti-symmetric character. Then by (6.5) $\chi = \psi\delta$, where ψ is a character. Now

$$(\text{sign}\,\varphi)\psi\delta = (\text{sign}\,\varphi)(\varphi\psi)\delta$$

and $\varphi\psi(t) = \psi(t)$ except, perhaps, where $\delta(t) = 0$, that is, on $\cup_r U_r$. Hence by continuity, $\varphi\psi(t) = \psi(t)$ for all $t \in T$ and ψ is symmetric. Thus the map is onto.

6.7 DEFINITION. Let $h \in L(T)^*$ be a weight, and let Wh be the orbit of h under W. Then the <u>elementary symmetric sum</u> $S(h)$ is given by

$$S(h) = \Sigma_{w \in Wh} \text{Exp}\ 2\pi i w.$$

6.8 EXAMPLE. Let $G = SU(n)$.

Then

$$S(x_1) = \xi_1 + \ldots + \xi_n,$$

where $\xi_j = \mathrm{Exp}\ 2\pi i x_j$

$$\begin{aligned} S(x_1 + x_2) = \xi_1 \xi_2 + \xi_1 \xi_3 & \ldots + \xi_1 \xi_n \\ + \xi_2 \xi_3 & \ldots + \xi_2 \xi_n \\ & \ldots \\ & + \xi_{n-1} \xi_n \end{aligned}$$

$$\begin{aligned} S(2x_1 + x_2) = \xi_1^2 \xi_2 + & \ldots \quad \xi_1^2 \xi_n \\ + \xi_2^2 \xi_1 + \xi_2^2 \xi_3 & \ldots \quad \xi_2^2 \xi_n \\ & \ldots \\ + \xi_n^2 \xi_1 & \ldots \quad \xi_n^2 \xi_{n-1} \end{aligned}$$

6.9 PROPOSITION. Let h run over a set of representatives of the orbits. Then $S(h)$ runs over a Z-basis for the symmetric elements of $K(T)$.

This is obvious.

6.10 EXAMPLE. In 6.9, h may run over the weights in Cl FDWC.

6.11 LEMMA. Let χ be an anti-symmetric character, and $h \in L(T)^*$ a singular weight, that is, $h \in L(U_r)^*$ for some r. Then $\mathrm{Exp}\ 2\pi i h$ occurs with coefficient 0 in χ.

Proof. Suppose

$$\chi = a \operatorname{Exp} 2\pi i h + \ldots.$$

Let $\varphi \in W$ be reflection in $L(U_r)^*$. Then

$$-\chi = \varphi\chi = a \operatorname{Exp} 2\pi i h + \ldots.$$

Thus $a = 0$.

6.12 **DEFINITION.** Let $h \in L(T)^*$ be a weight. Then the elementary alternating sum $A(h)$ is given by

$$A(h) = \Sigma_{\varphi \in W} (\operatorname{sign} \varphi) \operatorname{Exp} 2\pi i \varphi h.$$

If h is singular, then $A(h) = 0$. Otherwise, $A(h)$ contains $|W|$ distinct terms.

6.13 **EXAMPLE.** Let $G = SU(n)$, and let

$$h = a_1 x_1 + \ldots + a_{n-1} x_{n-1}.$$

Then

$$A(h) = \det \begin{bmatrix} \xi_1^{a_1} & \ldots & \xi_1^{a_{n-1}} & 1 \\ \xi_2^{a_1} & \ldots & \xi_2^{a_{n-1}} & 1 \\ & \ldots & & \\ \xi_n^{a_1} & \ldots & \xi_n^{a_{n-1}} & 1 \end{bmatrix}$$

6.14 **PROPOSITION.** Let h run over a set of representatives of orbits of regular weights. Then $A(h)$ runs over a Z-basis for the anti-symmetric characters.

This is obvious.

6.15 EXAMPLE. In 6.14, h may run over the weights in FDWC.

6.16 PROPOSITION. Let χ be the character of an irreducible complex representation of G, and let $\psi = \chi|T$. Then $\psi\delta = A(h)$ for some weight h.

Proof. If A_1, A_2 are elementary alternating sums, then

$$\begin{aligned}\int_T \bar{A}_1 A_2 &= |W| && \text{if } A_1 = A_2\\ &= -|W| && \text{if } A_1 = -A_2\\ &= 0 && \text{if } A_1 \neq \pm A_2\end{aligned}$$

by 3.34, since any weight is a character for T. Now $\psi\delta$ may be expressed as $\Sigma n_i A_i, n_i \in Z$ (6.14), so

$$1 = \int_G \bar{\chi}\chi = \frac{1}{|W|}\int_T \bar{\psi}\psi\bar{\delta}\delta$$

by 6.1

$$= \frac{1}{|W|}\int_T (\Sigma n_i \bar{A}_i)(\Sigma n_j A_n) = \Sigma n_i^2 .$$

Thus one n_i is ± 1 and the rest zero. By a suitable choice of h this can be made +1. Hence the result.

6.17 PROPOSITION. As χ runs over the characters of the distinct irreducible complex representations of G, the corresponding A(h) are all distinct.

Proof. If A_1, A_2 correspond to χ_1, χ_2 then

$$\frac{1}{|W|}\int_T \bar{A}_1 A_2 = \frac{1}{|W|}\int_T \bar{\psi}_1 \bar{\delta}\psi_2 \delta = \int_G \bar{\chi}_1 \chi_2 = 1 \quad \text{if } \chi_1 = \chi_2$$
$$= 0 \quad \text{if } \chi_1 \neq \chi_2 .$$

Hence the result.

To give a second proof, let $K(T)_W$ consist of the symmetric elements in K(T), that is, the elements invariant under W; and let $K(T)_{-W}$ consist of the antisymmetric elements. Consider the following composite.

$$K(G) \to K(T)_W \to K(T)_{-W} .$$

The first map is mono by 4.31, the second is iso by 6.6.

6.18 PROPOSITION. Suppose Ad lifts to Spin(n). Then every alternating sum A(h) arises as $\pm\psi\delta$ for some irreducible representation of G.

Proof. $A(h) = \sigma\delta$ for some symmetric character σ in K(T), and $\sigma = f|T$ for some class function f on G (4.32).

Now let χ be the character of an irreducible complex representation of G. Then

$$\int_G \bar{\chi} f = \frac{1}{|W|}\int_T \bar{\chi}\sigma\bar{\delta}\delta$$

$$= \frac{1}{|W|}\int_T \overline{A(k)}\,A(h), \quad \text{where } A(k) \text{ corresponds to } \chi$$

$$= 0 \quad \text{unless } A(k) = \pm A(h)$$

(see 6.16, proof). By the Peter-Weyl theorem (3.47) this is not zero for all χ. Hence $A(h) = \pm A(k)$ for some irreducible representation χ of G.

Summarising, we now have:

6.19 PROPOSITION. Suppose $\mathrm{Ad} : G \to SO(n)$ lifts to $\mathrm{Spin}(n)$. Then we have a 1-1 correspondence between irreducible representations of G and elementary alternating sums in $K(T)$, given by

$$K(G) \xrightarrow{\cong} K(T)_W \xrightarrow{\cong} K(T)_{-W} .$$

6.20 THEOREM. If G is compact and connected (but without the assumption that Ad lifts) then the map

$$K(G) \to K(T)_W$$

is an isomorphism.

Proof. The map is mono, by 4.31.

Now, if Ad does not lift to $\mathrm{Spin}(n)$, we have the following diagram (5.56(ii)):

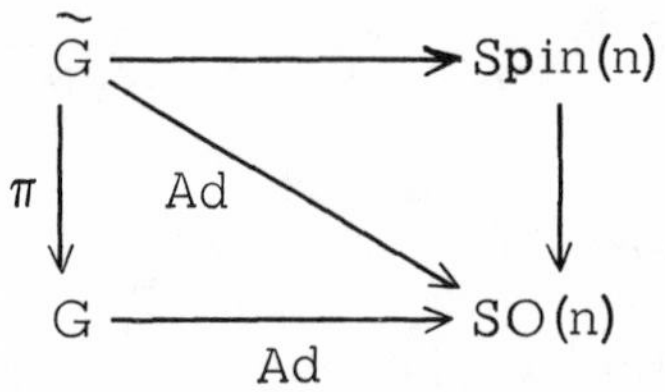

Let ψ be a symmetric element of $K(T)$. Then $\psi\pi \in K(\widetilde{T})$ is a symmetric element, so $\psi\pi = \widetilde{\chi}|\widetilde{T}$ for some virtual character $\widetilde{\chi}$ of $\widetilde{G}$ (6.19). Now $\widetilde{\chi}|\widetilde{T}$ factors through G so $\bar{\chi}$ factors through G (4.32). Thus (3.68) $\widetilde{\chi} = \chi\pi$ for some virtual character χ of G, and $\chi|T = \psi$.

6.21 REMARK. Even in the case where Ad does not lift to Spin, we can define anti-symmetric elements as those $\chi \in K(\widetilde{T})$ such that $\varphi\chi = (\text{sign}\,\varphi)\chi$ and $\chi(xz) = -\chi(x)$ for $1 \neq z \in \text{Ker}\,\pi$.

6.22 DISCUSSION. We are going to show that, when $\pi_1(G) = 0$, $K(G)$ is a polynomial algebra. Classically, the weights are ordered in a somewhat arbitrary way. When we propose to prove P = Q + "lower terms" the error must be "lower" with respect to all choices of ordering. So we will introduce an invariant partial order with which P = Q + "lower terms" will have approximately this meaning.

6.23 DEFINITION. Let ω_1, ω_2 be weights in $L(T)^*$. Define a partial order on the weights in $L(T)^*$ by writing $\omega_1 \leq \omega_2$ if ω_1 lies in the convex hull of the orbit of ω_2 under W. That is, $\omega_1 \leq \omega_2$ if $\omega_1 = \Sigma_{\varphi \in W} c_\varphi (\varphi \omega_2)$ for some non-negative coefficients c_φ with $\Sigma c_\varphi = 1$.

It is clear that $\omega_1 \leq \omega_2$ implies $\varphi_1 \omega_1 \leq \varphi_2 \omega_2$ for any $\varphi_1, \varphi_2 \in W$, that is, we are ordering the orbits. So it will suffice to consider weights in Cl FDWC.

6.24 **ALTERNATIVE DEFINITION.** For ω_1, ω_2 weights in Cl FDWC, write $\omega_1 \leq \omega_2$ if $\omega_1(v) \leq \omega_2(v)$ for all $v \in$ FWC. We may equally take all $v \in$ Cl FWC.

6.25 **PROPOSITION.** These two definitions are equivalent. We need:

6.26 **LEMMA.** If $u \in$ Cl FDWC, $v \in$ FWC and $\varphi \in W$, then $(\varphi u)(v) \leq u(v)$ with equality only if $\varphi u = u$.

Proof. If $\varphi u = u$ then $(\varphi u)(v) = u(v)$.

Suppose $\varphi u \neq u$ and $(\varphi u)(v) \geq u(v)$. Then, by moving v slightly in FWC, $(\varphi u)(v) > u(v)$.

Among the finite number of ψu as ψ runs through W, there is one, say ω, such that (ω, v) is maximal, and so $\omega \neq u$.

Then $\omega \notin \text{Cl FDWC}$ (5.16), so there is a simple root θ_r such that $\langle \theta_r, \omega \rangle < 0$. Consider $\varphi_r \omega$. We have

$$(\varphi_r \omega)(v) = \left(\omega - \frac{2\langle \theta_r, \omega \rangle}{\langle \theta_r, \theta_r \rangle} \theta_r \right)(v)$$

$$= \omega(v) - \frac{2\langle \theta_r, \omega \rangle}{\langle \theta_r, \theta_r \rangle} \theta_r(v)$$

$$> \omega(v),$$

which contradicts the definition of ω. So the result is proved.

The inequality $(\varphi u)v \leq u(v)$ remains true for $v \in \text{Cl FWC}$, by continuity.

Proof of 6.25

(i) If the first definition holds, then $\omega_1 = \Sigma c_\varphi (\varphi \omega_2)$ with $0 \leq c_\varphi, \Sigma c_\varphi = 1$, and we assume that $\omega_2 \in \text{Cl FDWC}$. Then for all $v \in \text{FWC}$ we have

$$\omega_1(v) = \Sigma c_\varphi (\varphi \omega_2)(v)$$

$$\leq \Sigma c_\varphi \omega_2(v) \quad \text{by } 6.26$$

$$= \omega_2(v).$$

So the second definition holds.

(ii) Suppose, on the contrary, that ω_1 does not lie in the convex hull of the orbit of ω_2, and suppose $\omega_1 \in \text{Cl FDWC}$. Then there is $\eta \in L(T)$ such that $\omega_1(\eta) > (\varphi \omega_2)(\eta)$ all $\varphi \in W$.

Write $\eta = \psi(v)$ with $\psi \in W$ and $v \in \mathrm{Cl}$ FWC. Then

$$\omega_1(v) \geq \omega_1(\eta) \quad (6.26)$$
$$> (\psi\omega_2)(\eta) = \omega_2(\psi\eta) = \omega_2(v),$$

so the second definition does not hold.

6.27 PROPERTIES OF THE RELATION

(i) Transitive: $\omega_1 \leq \omega_2 \leq \omega_3$ implies $\omega_1 \leq \omega_3$, obviously.

(ii) Given ω_2, the number of weights ω_1 such that $\omega_1 \leq \omega_2$ is finite. This is clear from the first definition.

[Note: This is better than the classical ordering, which allows one to make proofs by induction over the ordering only for the semi-simple Lie groups. For example, U(n) is not semi-simple.]

(iii) $\omega_1 \leq \omega_2$ and $\omega_2 \leq \omega_1$ if and only if $\omega_1 = \varphi\omega_2$ for some $\varphi \in W$, as follows. It suffices to consider $\omega_1, \omega_2 \in \mathrm{Cl}$ FDWC. If $\omega_1 \neq \omega_2$ we could find v in any open set of L(T) with $\omega_1(v) \neq \omega_2(v)$, contradicting the second definition.

6.28 DEFINITION. Write $\omega_1 < \omega_2$ if $\omega_1 \leq \omega_2$ but not $\omega_2 \leq \omega_1$. We then say that ω_1 is <u>lower</u> than ω_2.

6.29 EXERCISES. $u \leq v < w$ implies $u < w$ and $u < v \leq w$ implies $u < w$.

Continuation of 6.27

(iv) If $u, v, w \in$ Cl FDWC then $u + w \leq v + w$ if and only if $u \leq v$.

(v) If $t, u, v, w \in$ Cl FDWC and $t \leq u$, $v \leq w$ then $t + v \leq u + w$ with equality only if $t = u$ and $v = w$.

Let $\beta = \frac{1}{2}(\theta_1 + \ldots + \theta_m)$, and let ω be a weight in Cl FDWC.

6.30 PROPOSITION. If Ad lifts to Spin, then

$$S(\omega)\delta = A(\omega + \beta) + \text{lower terms},$$

that is,

$$S(\omega)\delta = A(\omega + \beta) + \Sigma n_i A(\omega_i)$$

with $\omega_i < \omega + \beta$.

Proof. (See 6.6.)

$$S(\omega) = \Sigma \operatorname{Exp}(2\pi i \omega_j),$$

where ω_j runs over the distinct $\varphi\omega$;

$$\delta = \Sigma \pm \operatorname{Exp}(2\pi i u_k),$$

where

$$u_k = \frac{1}{2}[\pm\theta_1 \ldots \pm\theta_m].$$

So

$$S(\omega)\delta = \Sigma \pm \operatorname{Exp} 2\pi i(\omega_j + u_k)$$

$$= \Sigma A(\omega_i),$$

where ω_i runs over those $\omega_j + u_k$ in Cl FDWC.

Now, if $x \in$ FWC, $\omega_j(x) = (\varphi\omega)(x) \leq \omega(x)$ with equality only if $\varphi\omega = \omega$ (6.26), and $u_k(x) \leq \beta(x)$ with equality only if $u_k = \beta$. Thus, if $\omega_j + u_k \in$ Cl FDWC, $\omega_j + u_k \leq \omega + \beta$ with equality only for the term $\omega_j = \omega$, $u_k = \beta$, which occurs with coefficient +1.

6.31 PROPOSITION. If Ad lifts to Spin, then

$$\frac{A(\omega + \beta)}{\delta} = S(\omega) + \text{lower terms}.$$

Proof. By induction. Suppose this is true for all $\omega' < \omega$. Then (6.30)

$$S(\omega)\delta = A(\omega + \beta) + \Sigma n_i A(\omega_i)$$

with $\omega_i < \omega$ and

$$\frac{A(\omega_i)}{\delta} = \Sigma m_{ij} S(\omega_j)$$

with $\omega_j \leq \omega_i$. So

$$\frac{A(\omega + \beta)}{\delta} = S(\omega) - \Sigma_{ij} n_i m_{ij} s(\omega_j)$$

with $\omega_j < \omega$.

6.32 EXAMPLE. If $\omega = 0$ we have $\frac{A(\beta)}{\delta} = S(0) = 1$. That is, $A(\beta) = \delta$.

6.33 **THEOREM**. There is a 1-1 correspondence between irreducible complex representations of G and weights ω in Cl FDWC in which $\chi|T = S(\omega)$ + lower terms.

Proof

(i) If Ad lifts to Spin, then $(\chi|T)\delta = A(\omega + \beta)$ sets up the correspondence (6.19 and 5.58) and

$$\chi|T = \frac{A(\omega + \beta)}{\delta} = \dot{S}(\omega) + \text{lower terms}$$

(6.31).

(ii) If Ad does not lift to Spin define $\pi : \tilde{G} \to G$ as before (5.56(ii)). For $\tilde{G}$, (i) holds.

For G we have $\chi|T = \Sigma n_i S(\omega_i)$, where ω_i runs over weights of G (6.9), so $\chi\pi|\tilde{T} = \Sigma n_i S(\omega_i)$, ω_i being interpreted as weights of $\tilde{G}$. If χ is irreducible then $\chi\pi$ is irreducible, so $\chi\pi|\tilde{T} = S(\omega)$ + lower terms. Therefore, by the uniqueness of such expressions, $\chi|T = S(\omega)$ + lower terms. This sets up the correspondence and shows that it is mono.

Now let ω be a weight for G in Cl FDWC, and let $\tilde{\chi}$ be a character of $\tilde{G}$ such that $\tilde{\chi}|T = \dfrac{A(\omega + \beta)}{\delta}$. Then $\tilde{\chi}|\tilde{T}$ factors through T, so χ factors through G both as a function and (3.68) as a character. Since $\tilde{\chi}$ is irreducible, so is χ.

6.34 DEFINITION. It follows that each irreducible representation of G has associated a <u>maximal weight</u>, which occurs with multiplicity one.

6.35 EXAMPLE. Let $G = SU(n)$.

For $\omega = x_1$ we have

$$\omega + \beta = nx_1 + (n-2)x_2 + \ldots + x_{n-1}.$$

Then

$$\frac{A(\omega+\beta)}{\delta} = \frac{A(\omega+\beta)}{A(\beta)} = \frac{\begin{vmatrix} \xi_1^n & \xi_1^{n-2} & \ldots & \xi_1 & 1 \\ \xi_2^n & \xi_2^{n-2} & \ldots & \xi_2 & 1 \\ & & \ldots & & \end{vmatrix}}{\begin{vmatrix} \xi_1^{n-1} & \xi_1^{n-2} & \ldots & \xi_1 & 1 \\ \xi_2^{n-1} & \xi_2^{n-2} & \ldots & \xi_2 & 1 \\ & & \ldots & & \end{vmatrix}}$$

$$= \xi_1 + \ldots + \xi_n.$$

For $\omega = 2x_1$ we get

$$\Sigma \xi_i^2 + \sum_{i<j} \xi_i \xi_j.$$

6.36 PROPOSITION. Let u,v be weights in Cl FDWC. Then

$$S(u)S(v) = S(u+v) + \text{lower terms}.$$

Proof. Let

$$S(u) = \Sigma \operatorname{Exp} 2\pi i u_j, \quad S(v) = \Sigma \operatorname{Exp} 2\pi i v_k,$$

where u_j, v_k run over the distinct $\varphi u, \varphi v$ for $\varphi \in W$. Then

$$S(u)S(v) = \Sigma \operatorname{Exp} 2\pi i(u_j + v_k).$$

If $x \in$ FWC, then (6.26) $(\varphi u)(x) \leq u(x)$ and $(\varphi v)(x) \leq v(x)$ with equality holding if $\varphi u = u$, $\varphi v = v$ respectively. That is,

$$(u_j + v_k)(x) \leq (u + v)(x)$$

with equality only for the single term $u_j = u$, $v_j = v$. Thus, if $u_j = v_k \in$ Cl FDWC, then $u_j + v_k < u + v$ except in the single case $u_j = u$, $v_k = v$. This gives the result.

6.37 EXAMPLES. Let $G = SU(n)$ (see 6.8).

$$(\Sigma \xi_i)(\Sigma \xi_j) = \Sigma \xi_i^2 + 2 \sum_{i<j} \xi_i \xi_j$$

$$= \Sigma \xi_i^2 + \text{lower terms}.$$

$$(\Sigma \xi_i)\Big(\sum_{i<k} \xi_i \xi_k\Big) = \sum_{i \neq j} \xi_i^2 \xi_j + 3 \Sigma_{i<j<k} \xi_i \xi_j \xi_k$$

$$= \sum_{i \neq j} \xi_i^2 \xi_j + \text{lower terms}.$$

Let $G = SO(2n + 1)$.

$$(\Sigma \xi_i + \Sigma \xi_i^{-1})(\Sigma \xi_i + \Sigma \xi_i^{-1})$$

$$= \left(\Sigma \xi_i^2 + \Sigma \xi_i^{-2}\right) + 2\left(\sum_{i<j} \xi_i \xi_j + \sum_{i \neq j} \xi_i \xi_j^{-1} + \sum_{i<j} \xi_i^{-1} \xi_j^{-1}\right) + 2n$$

$$= \left(\Sigma \xi_i^2 + \Sigma \xi_i^{-2}\right) + \text{lower terms}.$$

6.38 DISCUSSION. Suppose $\pi_1(G) = 0$. We know that the weights in Cl FDWC form a free semi-group generated by $\omega_1, \ldots, \omega_k$ (5.62). So there are irreducible representations $\rho_1, \ldots, \rho_k$ of G such that

$$\chi(\rho_r)|T = S(\omega_r) + \text{lower terms}.$$

Using 6.36 inductively,

$$\chi\left(\rho_1^{n_1} \ldots \rho_k^{n_k}\right)|T = S(n_1\omega_1 + \ldots + n_k\omega_k) + \text{lower terms}.$$

6.39 PROPOSITION. If $\pi_1(G) = 0$, then

$$Z[\rho_1, \ldots, \rho_k] \to K(G)$$

is mono.

Proof. Let

$$a_1 m_1 + \ldots + a_r m_r = 0$$

be a linear combination of distinct monomials m_i in the ρ's with $0 \neq a_i \in Z$. Since the monomials are in 1-1 correspondence with the weights in Cl FDWC, we can order the monomials by reference to the weights. If the linear combination is non-empty, then it contains an m_i such that no $m_j > m_i$. Let ω be the weight corresponding to m_i. Then in

$$\chi(a_1 m_1 + \ldots + a_r m_r)|T$$

the only term in $S(\omega)$ is $a_i S(\omega)$, so $a_i = 0$, which is a contradiction.

6.40 PROPOSITION

$$S(n_1\omega_1 + \ldots + n_k\omega_k) = \chi\left(\rho^{n_1} \ldots \rho_k^{n_k} + \text{lower monomials}\right) | T.$$

Proof. We proceed by induction. Write

$$\omega = n_1\omega_1 + \ldots + n_k\omega_k,$$

and suppose the result is true for all $\omega' < \omega$.

$$\chi\left(\rho_1^{n_1} \ldots \rho_k^{n_k}\right) | T = S(\omega) + \Sigma m_i S(\omega_i),$$

where $\omega_i < \omega$. By the induction hypothesis,

$$S(\omega_i) = \chi(\text{lower monomials}) | T.$$

Therefore

$$S(\omega) = \chi\left(\rho_1^{n_1} \ldots \rho_k^{n_k} + \text{lower monomials}\right) | T.$$

6.41 THEOREM. Let G be a compact connected simply-connected Lie group. Then

$$K(G) \cong Z[\rho_1, \rho_2, \ldots, \rho_k].$$

Proof. By 6.39,

$$Z[\rho_1, \ldots, \rho_k] \to K(G)$$

is mono. By 6.40, the following composite is epi:

$$Z[\rho_1, \ldots, \rho_k] \to K(G) \xrightarrow{\cong} K(T)_W.$$

So the map is iso.

Chapter 7

REPRESENTATIONS OF THE CLASSICAL GROUPS

In this chapter we will derive the complex representation rings of the classical compact Lie groups. We will also enquire if each group has any irreducible representations which are real or quaternionic. For this purpose we consider the following maps:

$$K(G) \xrightarrow{1+t} K(G) \xrightarrow{1-t} K(G).$$

We define

$$H = \mathrm{Ker}(1 - t)/\mathrm{Im}(1 + t).$$

7.1 **PROPOSITION.** H is an algebra over Z_2, and the irreducible representations of G which are self-conjugate yield a Z_2-base for H.

The proof is immediate from Chapter 3. We may therefore measure the incidence of self-conjugate irreducible representations by computing H.

We will also use the following lemma.

7.2 LEMMA For any complex representation V, $V^* \otimes V \cong \mathrm{Hom}(V,V)$ is real.

Proof. It carries the bilinear form

$$\mathrm{Tr}(\alpha\beta) = \mathrm{Tr}(\beta\alpha)$$

(see 3.38); this form is symmetric, non-singular and invariant. Now use 3.50.

We now begin to study the groups $U(n)$ and $SU(n)$. Each has an obvious representation with $V = C^n$; we write $\lambda^1, \lambda^2, \ldots, \lambda^n$ for the exterior powers of this operation. Let us write

$$z_j = \mathrm{Exp}(2\pi i x_j),$$

so that the typical element in our maximal torus is

$$D = \begin{bmatrix} z_1 & & & \\ & z_2 & & \\ & & \ddots & \\ & & & z_n \end{bmatrix};$$

then the character $\chi(\lambda^k)$ of λ^k is the kth elementary symmetric function of $z_1, z_2, \ldots, z_n$. (See the proof of 3.61.) The Weyl group acts by permuting $z_1, z_2, \ldots, z_n$. Thus the character $\chi(\lambda^k)$ is the elementary symmetric sum

$$S(x_1 + x_2 + \ldots + x_k).$$

Since $\chi(\lambda^k)$ consists of a single elementary symmetric sum. λ^k is irreducible. The representation λ^n of $U(n)$ is one-dimensional, and is essentially det : $U(n) \to S^1$. In particular, it is invertible. The restriction of λ^n to $SU(n)$ is trivial.

7.3 **THEOREM.** The complex representation ring $K(U(n))$ is the tensor product of the polynomial ring generated by $\lambda^1, \lambda^2, \ldots, \lambda^{n-1}$ and the ring of finite Laurent series in λ^n. The algebra H is polynomial on generators $\lambda^i \lambda^{n-i}/\lambda^n$ for $2 \leq 2i \leq n$. These generators are real.

There is of course no suggestion that the modules $\lambda^i \lambda^{n-i}/\lambda^n$ are irreducible; indeed they are not.

Proof of 7.3. By a classical theorem, the ring of symmetric polynomials in $z_1, z_2, \ldots, z_n$ is a polynomial ring generated by the elementary symmetric functions $\chi(\lambda^1), \ldots, \chi(\lambda^n)$. Now take any finite Laurent series which is symmetric; by multiplying it with a suitably high power of $z_1 z_2 \ldots z_n$, we obtain a symmetric polynomial. Hence $K(T)_W$ is as described.

The result for K(U(n)) follows by 6.20.

Since we have an obvious pairing

$$\lambda^i \otimes \lambda^{n-i} \to \lambda^n,$$

the dual of λ^i is λ^{n-i}/λ^n. (This also follows from an easy calculation with characters.) Hence the conjugate of

$$(\lambda^1)^{\nu_1}(\lambda^2)^{\nu_2} \ldots (\lambda_n)^{\nu_n}$$

is

$$(\lambda^1)^{\nu_{n-1}}(\lambda^2)^{\nu_{n-2}} \ldots (\lambda^{n-1})^{\nu_1}(\lambda^n)^{-\nu_1-\nu_2\ldots-\nu_n}.$$

So t permutes the monomials in $\lambda^1, \lambda^2, \ldots, \lambda^n$; and we easily see that the only monomials which are fixed under t are the polynomials in

$$\lambda^i \lambda^{n-i}/\lambda^n \qquad (1 \leq i \leq \tfrac{1}{2}n).$$

These are real by 7.2, since

$$\mathrm{Hom}(\lambda^i, \lambda^i) \cong \lambda^i \lambda^{n-i}/\lambda^n.$$

7.4 THEOREM. The complex representation ring K(SU(n)) is a polynomial ring generated by $\lambda^1, \lambda^2, \ldots, \lambda^{n-1}$. The algebra H is polynomial on generators $\lambda^i\lambda^{n-i}$ for $2 \leq 2i < n$ and, if $n = 2m$, a generator λ^m. The generators $\lambda^i\lambda^{n-i}$ are real; the generator λ^m is real for m even, quaternionic for m odd.

Proof. The result on K(SU(n)) is a special case of 6.41;

the identification of the basic weights $\omega_1, \ldots, \omega_k$ mentioned in 6.38 is given in 5.63.

As above, the dual of λ^i is λ^{n-i}. Hence the conjugate of

$$(\lambda^1)^{\nu_1}(\lambda^2)^{\nu_2} \ldots (\lambda^{n-1})^{\nu_{n-1}}$$

is

$$(\lambda^1)^{\nu_{n-1}}(\lambda^2)^{\nu_{n-2}} \ldots (\lambda^{n-1})^{\nu_1}.$$

So t permutes the monomials in $\lambda^1, \lambda^2, \ldots, \lambda^{n-1}$; and we easily see that the only monomials which are fixed under t are polynomials in $\lambda^i\lambda^{n-i}$ $(1 \leq i < \frac{1}{2}n)$ and λ^m if $n = 2m$. The representation $\lambda^i\lambda^{n-i}$ is real by 7.2. As for λ^m, the pairing

$$\lambda^m \otimes \lambda^m \to \lambda^{2m} = C$$

has

$$\beta \wedge \alpha = (-1)^m \alpha \wedge \beta ;$$

now use 3.50.

7.5 **EXERCISE.** Show directly that any representation V of SU(n) extends to U(n). (Hint: It is sufficient to consider an irreducible representation; now consider the action of the centre of SU(n).)

We take next the group Sp(n). It has an obvious representation on $Q^n \cong C^{2n}$; we write $\lambda^1, \lambda^2, \ldots, \lambda^{2n}$ for the exterior powers of this representation. As we have seen in

Chapter 3, λ^k is real for k even, quaternionic for k odd. If we take the element

$$D = \begin{bmatrix} z_1 & & & \\ & z_2 & & \\ & & \ddots & \\ & & & z_n \end{bmatrix}$$

in T, its action on C^{2n} is given by

$$\begin{bmatrix} z_1 & & & & & & \\ & \bar{z}_1 & & & & & \\ & & z_2 & & & & \\ & & & \bar{z}_2 & & & \\ & & & & \ddots & & \\ & & & & & z_n & \\ & & & & & & \bar{z}_n \end{bmatrix}.$$

Therefore the character $\chi(\lambda^i)$ of λ^i is the ith elementary symmetric function of

$$z_1, z_1^{-1}, z_2, z_2^{-1}, \ldots, z_n, z_n^{-1}.$$

7.6 THEOREM. K(Sp(n)) is a polynomial algebra with generators $\lambda^1, \lambda^2, \ldots, \lambda^n$. All the irreducible representations of Sp(n) are self-conjugate.

Proof

(i) It is rather easy to see that $K(T)_W$ is as stated; now use 6.20. Alternatively, use 6.41.

(ii) It follows from the generators given that the whole of K(Sp(n)) is self-conjugate. Alternatively, in Sp(n) each element g is conjugate to g^{-1} (see 5.17).

We take next the group SO(n). It has an obvious representation on R^n or C^n; we write $\lambda^1, \lambda^2, \ldots, \lambda^n$ for the exterior powers of this representation. All these representations are real. If we take the element

$$D = \begin{bmatrix} z_1 & & & \\ & z_2 & & \\ & & \ddots & \\ & & & z_n \end{bmatrix}$$

in U(n) and embed it in SO(2n), its action on C^{2n} is equivalent to that of the diagonal matrix

$$\begin{bmatrix} z_1 & & & & & & \\ & \bar{z}_1 & & & & & \\ & & z_2 & & & & \\ & & & \bar{z}_2 & & & \\ & & & & \ddots & & \\ & & & & & z_n & \\ & & & & & & \bar{z}_n \end{bmatrix}.$$

Therefore the character $\chi(\lambda^i)$ of λ^i is the ith elementary symmetric function of

$$z_1, z_1^{-1}, z_2, z_2^{-1}, \ldots, z_n, z_n^{-1},$$

say σ_i. Similarly, if we embed D in SO(2n + 1), its action on C^{2n+1} is equivalent to that of the diagonal matrix

$$\begin{bmatrix} z_1 & & & & & \\ & \bar{z}_1 & & & & \\ & & \ddots & & & \\ & & & z_n & & \\ & & & & \bar{z}_n & \\ & & & & & 1 \end{bmatrix}.$$

Therefore we have

$$\chi(\lambda^i) = \sigma_i + \sigma_{i-1}.$$

(Here σ_0 is to be interpreted as 1.)

7.7 THEOREM. K(SO(2n + 1)) is a polynomial algebra with generators $\lambda^1, \lambda^2, \ldots, \lambda^n$. All the irreducible representations of SO(2n + 1) are real.

Proof

(i) $K(T)_W$ is exactly the same as for Sp(n).

(ii) It follows from the generators given that the whole of K(SO(2n +1)) is real.

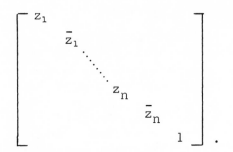

So far the exterior powers λ^i have given us all the generators we need. It is easy to produce arguments to show that for SO(2n) we need something else.

(i) In SO(4n + 2) not every element g is conjugate to g^{-1} Therefore it is possible to construct a class function f such that $f(g) \neq f(g^{-1})$. Therefore (3.47) SO(4n + 2) has at least one representation which is not self-conjugate. But all the λ^i are real.

(ii) Consider the representation λ^n of SU(2n). We have already seen that it is self-conjugate. So its restriction to SO(2n) is self-conjugate for two essentially different reasons: first because λ^n is self-conjugate on SU(2n), and secondly because each exterior power λ^i is real on SO(2n). But we have already seen that an irreducible representation V can have essentially only one isomorphism with V*. Therefore the representation λ^n of SO(2n) is reducible.

If n is odd this argument is complete in itself; the representation λ^n of SO(2n) is both quaternionic and real, so it cannot be irreducible. If n is even it is desirable to amplify the word "essentially" a little, and this will be done below.

(iii) An alternative argument proceeds by considering the representation λ^n of O(2n). Consider an element g in O(2n) such that $\det(g) = -1$; it is easy to see that its action on C^{2n} is equivalent to that of a diagonal matrix

$$\begin{bmatrix} z_1 & & & & & & & \\ & \bar{z}_1 & & & & & & \\ & & \ddots & & & & & \\ & & & z_{n-1} & & & & \\ & & & & \bar{z}_{n-1} & & & \\ & & & & & 1 & & \\ & & & & & & -1 \end{bmatrix}$$

It is now easy to check that the restriction of $\chi = \chi(\lambda^n)$ to the component of determinant -1 in $O(2n)$ is zero. Let the average value of $\bar{\chi}\chi$ over $SO(2n)$ be ν; then the average value of $\bar{\chi}\chi$ over $O(2n)$ is $\frac{1}{2}\nu$. So $\frac{1}{2}\nu \geq 1$ (3.34) and $\nu \geq 2$. That is, λ^n must split over $SO(2n)$ into at least two summands.

We now amplify argument (ii). Let us define a non-singular bilinear pairing

$$F : \lambda^n(R^{2n}) \otimes \lambda^n(R^{2n}) \to \lambda^{2n}(R^{2n}) = R$$

by $F(v,w) = v \wedge w$. Then F is invariant under $SO(2n)$; indeed for $g \in O(2n)$ we have

$$F(gv,gw) = (\det g)F(v,w).$$

Let us define another non-singular bilinear pairing

$$S : \lambda^n(R^{2n}) \otimes \lambda^n(R^{2n}) \to R$$

by

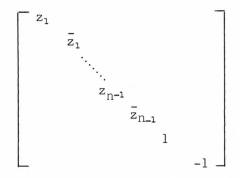

$$S\Big((v_1 \wedge v_2 \wedge \ldots \wedge v_n) \otimes (w_1 \wedge w_2 \wedge \ldots \wedge w_n)\Big)$$
$$= \sum_\rho \epsilon(\rho)(v'_{\rho(1)}w_1) \ldots (v'_{\rho(n)}w_n).$$

Here ρ runs over all permutations, and $v'w$ is the usual inner product in R^{2n}. Then S is invariant under $O(2n)$. Let us define an automorphism β of $\lambda^n(R^{2n})$ by setting

$$S(\beta v, w) = F(v, w).$$

We easily check that for $g \in O(2n)$ we have

$$\beta g v = (\det g)\beta v.$$

We may describe β as follows. Let $v_1, v_2, \ldots, v_{2n}$ be any orthonormal basis with determinant +1 in R^{2n}; then

$$\beta(v_1 \wedge v_2 \wedge \ldots \wedge v_n) = v_{n+1} \wedge v_{n+2} \wedge \ldots \wedge v_{2n}.$$

Thus $\beta^2 = (-1)^n$.

It follows that $\lambda^n(R^{2n})$ splits into the ± 1 eigenspaces of β if n is even, and into the $\pm i$ eigenspaces of β if n is odd. Of course the latter splitting takes place over C. Elements of $SO(2n)$ preserve the two eigenspaces; elements of determinant -1 in $O(2n)$ interchange the two eigenspaces. In particular, neither eigenspace can be zero.

We now enquire after the characters of the summands (say V and W). The character $\chi(\lambda^n)$ of λ^n is the nth elementary symmetric function of

$$z_1, z_1^{-1}, z_2, z_2^{-1}, \ldots, z_n, z_n^{-1}.$$

Let us write

$$\alpha_+ = \Sigma z_1^{\epsilon_1} z_2^{\epsilon_2} \ldots z_n^{\epsilon_n} \mid \epsilon_r = \pm 1 \text{ and } \epsilon_1 \epsilon_2 \ldots \epsilon_n = +1.$$

$$\alpha_- = \Sigma z_1^{\epsilon_1} z_2^{\epsilon_2} \ldots z_n^{\epsilon_n} \mid \epsilon_r = \pm 1 \text{ and } \epsilon_1 \epsilon_2 \ldots \epsilon_n = -1.$$

These are elementary symmetric sums (see 5.17). We have

$$\chi(\lambda^n) = \alpha_+ + \alpha_- + \text{lower terms}.$$

Since the characters of representations are linear combinations with non-negative coefficients of elementary symmetric sums, we have

$$\chi(V) = a\alpha_+ + b\alpha_- + \sigma$$

where a and b are 0 or 1, and σ is a sum of lower terms. Now consider the automorphism θ of SO(2n) obtained by conjugating with an element g of determinant -1 in O(2n), say

$$g = \begin{bmatrix} 1 & & & & \\ & 1 & & & \\ & & \ddots & & \\ & & & 1 & \\ & & & & -1 \end{bmatrix}.$$

Its effect on T is to invert z_n; thus $\theta\alpha_+ = \alpha_-$, $\theta\alpha_- = \alpha_+$ and $\theta\sigma = \sigma$. Hence

$$\chi(W) = b\alpha_+ + a\alpha_- + \sigma,$$

$$\chi(\lambda^n) = (a + b)(\alpha_+ + \alpha_-) + 2\sigma$$

and $a + b = 1$. It follows that we can name the summands of λ^n so that

$$\chi(\lambda^n_+) = \alpha_+ + \sigma$$

$$\chi(\lambda^n_-) = \alpha_- + \sigma .$$

7.8 COROLLARY. The automorphism θ of $SO(2n)$ is not inner.

Proof. An inner automorphism takes a representation into an equivalent representation.

7.9 THEOREM. $K(SO(2n))$ is a free module over the polynomial ring $Z[\lambda^1, \lambda^2, \ldots, \lambda^n]$ on two generators 1 and λ^n_+ (or equivalently 1 and λ^n_-). If n is even all the irreducible representations of $SO(2n)$ are real. If n is odd,

$$H = Z_2[\lambda^1, \lambda^2, \ldots, \lambda^{n-1}].$$

Proof. We have to study $K(T)_W$, that is, the set of finite Laurent series in $z_1, z_2, \ldots, z_n$ which are symmetric under permutations and under inverting an even number of the z_r. The set S of such symmetric elements admits an automorphism θ: invert an odd number of the z_r. (Of course, θ arises as explained above.) We have $\theta^2 = 1$. So over the rationals, S

splits as the sum of the $+1$ and -1 eigenspaces of θ:

$$s = \frac{1}{2}(1+\theta)s + \frac{1}{2}(1-\theta)s.$$

The $+1$ eigenspace is the ring of polynomials in

$$\chi(\lambda^1), \chi(\lambda^2), \ldots, \chi(\lambda^n)$$

(as in 7.6, 7.7). Suppose given an element $\underline{a}$ in the -1 eigenspace. Then

$$a = \sum_r c_r(z_1, z_2, \ldots, z_{n-1}) z_n^r,$$

where

$$c_{-r} = -c_r,$$

so that

$$a = \sum_1^n c_r(z_1, z_2, \ldots, z_{n-1})(z_n^r - z_n^{-r}).$$

Thus

$$a = a'(z_n - z_n^{-1}).$$

By symmetry $\underline{a}$ is divisible by the remaining $(z_r - z_r^{-1})$; so

$$a = a''(z_1 - z_1^{-1})(z_2 - z_2^{-1}) \ldots (z_n - z_n^{-1}).$$

Here a'' must be an element of the $+1$ eigenspace; so we have

$$a = p(\alpha_+ - \alpha_-)$$

where p is a polynomial in $\chi(\lambda^1), \ldots, \chi(\lambda^n)$. For a general element s in S we have

$$s = \frac{1}{2}(1+\theta)s + \frac{1}{2}p(\alpha_+ - \alpha_-).$$

Since $\alpha_+ + \alpha_-$ lies in the $+1$ eigenspace we may write this

$$s = q + p\alpha_+$$

where q lies in the $+1$ eigenspace and is integral (since s and $p\alpha_+$ are so).

So $K(T)_W$ is as claimed, and the result on $K(SO(2n))$ follows by 6.20.

If n is even all the generators for $K(SO(2n))$ are real. If n is odd $t(\lambda_+^n) = \lambda_-^n$, and the calculation of H is easy. This completes the proof.

For lack of time I have not included anything on the representation-theory of Spin(n). Of course, this is included as a special case of 6.41; but some may prefer to see the basic representations arise more directly. I advise such readers to study Clifford algebras out of [1] and the representations of the Clifford algebras out of [7]. In [7] Eckmann actually studies the representations of a certain finite group G, but the Clifford algebra is an obvious quotient of the group ring R(G), and so the representations of the Clifford algebra are easily read off from the representations of G.

REFERENCES

[1] M. F. Atiyah, R. Bott and A. Shapiro. Clifford modules, Topology, **3** Supplement 1 (1964), pp. 3-38.

[2] Bliss. Lectures on Calculus of Variations. University of Chicago Press (1946).

[3] A. Borel and F. Hirzebruch. Characteristic classes and homogeneous spaces I, Amer. J. Math., **80** (1958), pp. 458-538.

[4] R. Bott. Lectures on K(X). Mimeographed notes, Harvard University.

[5] Coddington and Levinson. Theory of Ordinary Differential Equations. McGraw-Hill (1955).

[6] A. Dold. Fixed point index and fixed point theorem for Euclidean neighbourhood retracts. Topology, **4** (1965), pp. 1-8.

[7] B. Eckmann. Gruppentheoretischer Beweis des Satzes von Hurwitz-Radon uber die Komposition quadratischer Formen. Comment.Math.Helv., **15** (1942), pp. 358-36

[8] Graves. Theory of Functions of Real Variables. McGraw-Hill (1946).

[9] G. Hochschild. The Structure of Lie Groups. Holden-Day (1965).

[10] H. Hopf. Maximale Toroide und singulare Elemente in geschlossen Lieschen Gruppe. Comment.Math.Helv., **15** (1943), pp. 59-70.

[11] H. Hopf and H. Samelson. Ein Satze uber die Wirkungsraume geschlossener Liescher Gruppen. Comment.Math.Helv., **13** (1941), pp. 240-251.

[12] S. Lang. Introduction to Differentiable Manifolds. Interscience (1962).

[13] Loomis. An Introduction to Abstract Harmonic Analysis. Van Nostrand (1953).

[14] L. Nachbin. The Haar Integral. Van Nostrand (1965).

[15] F. Peter and H. Weyl. Die Vollstandigkeit der primitive Darstelling einer geschlossenen kontinuerlichen Gruppe. Math. Ann., **97** (1927), pp. 737-755.

[16] F. Smithies. Integral Equations. Cambridge Univ. Press (1962).

[17] N. E. Steenrod. The Topology of Fibre Bundles. Princeton Univ. Press (1951).

[18] E. Stiefel. Uber eine Beziehung zwischen geschlossenen Lieschen Gruppen und diskontinuerlichen Bewegungsgruppen euklidischer Raume und ihrer Anwendung auf die Aufzahlung der einfachen Lieschen Gruppen. Comment.Math.Helv., **14** (1942), pp. 350-380.

[19] E. Stiefel. Kristallographische Bestimmung der Charaktere der geschlossenen Lieschen Gruppen. Comment. Math. Helv., **17** (1945), pp. 165-200.

[20] A. Weil. L'Integration dans les groupes topologiques. Hermann (1940).

[21] A. Weil. Demonstration topologique d'un théorème fondamental de Cartan. C. R. Acad. Sci., Paris, **200** (1935), pp. 518-520.

[22] H. Weyl. Theorie der Darstellung kontinuerlicher halb-einfacher Gruppen durch linearer Transformationen. Math. Zeit., **23** (1924), pp. 271-309; **24** (1925), pp. 328-376; **24** (1926), pp. 377-395.

[23] H. Weyl. The Classical Groups. Princeton Univ. Press (1946).